Repatriation Management and Competency Transfer in a Culturally Dynamic World

Manmeet Bali Nag · Firdous Ahmad Malik

Repatriation Management and Competency Transfer in a Culturally Dynamic World

Springer

Manmeet Bali Nag
Department of Management and Commerce
Manav Rachna University
Faridabad, Haryana, India

Firdous Ahmad Malik
Jindal Center for Global South
O. P. Jindal Global University
Sonipat, Haryana, India

ISBN 978-981-19-7352-9 ISBN 978-981-19-7350-5 (eBook)
https://doi.org/10.1007/978-981-19-7350-5

This Springer imprint is published by the registered company Springer Nature Singapore Pte Ltd.
The registered company address is: 152 Beach Road, #21-01/04 Gateway East, Singapore 189721, Singapore

Dedicated
To Our Revered
"Parents"

Acknowledgements

No one who achieves success does so without acknowledging the help of others. The wise and confident acknowledge this help with gratitude

—Alfred North Whitehead

This book is a labor of deeper research into the unchartered territory of Repatriation Management in Corporate Sector and its antecedent. Needless to say, it brings forth immense sense of satisfaction, accomplishment, and gratitude to all those who enabled it.

Most of the results deliberated upon in this thesis were obtained with a close collaboration with the corporate enterprise in North India. I owe a great deal of appreciation and gratitude to the HR personnel and the Repatriates from various organizations within the ambit of the Indian Information Technology Sector, e.g., Sapient Technologies, HCL, Informatica, Accenture, Aricent, Xerox, I-Gate, Amdocs, Infosys, Tech Mahindra, TCS, Wipro, Hexaware, TCS, Infosys, etc., to name a few, who by virtue of being the respondents filled the research questionnaire, despite their hectic schedules.

I extend my thanks to various sources and authors, mentioned in the bibliography, for facilitating me with the vast expanse of requisite knowledge. I acknowledge the role of my organization, Manav Rachna University, Faridabad, and the Department of Management and Commerce, including both my superiors and peers, for their steadfast support and insightful suggestions. This paved the way for shaping up my ideas and subsequently their implementation in my research.

Heartfelt thanks to my family, including my parents, husband, and son, for their support and unwavering belief in me that has sustained me this far, in all the endeavors of life. My earnest thanks to all my friends and well-wishers. Along the journey, I owe my gratitude to our editor Bartonek, Lucie, and Springer Group for providing me with this opportunity. I look forward to rewarding association in the future too. It has been a tremendous learning experience, encompassing various bottlenecks and accomplishments holistically, wherein I learned and unlearned from both aspects and thereby imbibed the requisite academic/research skills. The subsequent learning and evolution shall be cherished forever.

Manmeet Bali Nag

Prologue

The need for a global strategic perspective is gaining momentum, as international competition continues to intensify. On a global scale, one of the greatest challenges in the contemporary corporate scenario is how to manage human resources. A global business strategy requires globally competent people and global human resource systems, which need to be implemented successfully. Therefore, the human resource personnel need to maximize the potential of each employee, to harness the multi-faceted dimensions of their employees. As a result, International Human Resource Management has evolved a great deal in the last decade, with cross-cultural interactions being reflected in the form of business interpretations, and numerous studies being undertaken on expatriation. However, the theme of cross-cultural reentry (its course, impact, and features) still remains largely neglected and underestimated in the sojourner's transition trajectory. At the same time, available empirical investigations point to a number of substantial concerns, which affect the psychological well-being, social readjustment, and cultural identity of the returning individuals. The significance of these concerns strongly suggests that reentry should become an issue of the highest priority to both sojourning individuals and people managing the reentry transitions of travelers.

Presently, India is internationally known as an IT destination and is admired for its well-trained and educated human capital. The Information Technology Industry in India has transformed "Brand India". A knowledge revolution in the world and India has led to the participation of many countries in it, because of the high knowledge density and rich demographic dividend that the country offers. Therefore, it is for companies, to harness this knowledge capital and accelerate Indians in the knowledge revolution. As an attractive method for accumulating foreign markets and capital, organizations perceive expatriation as one of the best ways. But the optimization and management of this valuable resource in all its manifestations, namely expatriation as well as repatriation, still remain a challenge. This implies that even if an individual is sent for an international assignment, the problems are at times faced by individuals when they come back from the assignment to the home country. There are various factors that can influence the repatriation of an individual when back in the home country. A focus on this aspect of the international assignment therefore attains significance to leverage the human resource talent, lest it goes waste and consequently results in overall wastage and losses, from an organizational point of view.

While determining the evaluation of an international assignment, the impact of the strategic HR implementation of programs on repatriate retention, satisfaction, and job commitment or outcome measures and the participants' assessment of the effectiveness of different strategies or process evaluation have to be taken into account. The identification of gaps and continuous benchmarking of the overall strategy calls for a holistic research and analysis in the said domain of International Human Resource Management. The purview of this research encompasses the interface of Anticipatory Adjustment, Competency Transfer, Effective Repatriation, and Reverse Culture Shock with Repatriates' Adjustment, back home, and empirically analyzes the precursors and effects of the said paradigms in optimization of repatriate talent from an organizational perspective. It concentrates on the Repatriation Adjustment of Indian international assignees from Information Technology sector. The purpose of this research is to analyze the antecedents of repatriate adjustment in the mentioned context and to subsequently deliberate upon the acquired results to arrive at feasible and relevant conclusions.

Chapter 1: **Introduction**: This chapter contains an overview of Repatriation Management in the context of a globalized world and from International Human Resource Management (IHRM) perspective, followed by the importance of repatriate talent and its optimization. It serves as an introduction to the field of Repatriation Acculturation and Adjustment Antecedents and formalization of what knowledge the research aims to contribute. Repatriation as a strategic Human Resource Management tool for retaining manpower and talent has been highlighted, along with the various nuances which effect and channelize such talent in a particular direction, and has been deliberated upon.

Chapter 2: **Conceptual and Theoretical Background**: The second chapter focuses on the review of relevant literature. It contains an expanded discussion on the literature with a complete and comprehensive review of the various research studies that have been carried out in the context of the current study. The purpose of this chapter is to build a theoretical framework for the research by reviewing existing literature. It provides a thorough review of literature related to the Repatriation Adjustment and management, as an intertwined concept relying heavily on associated parameters and domains of Anticipatory Adjustment, Effective Repatriation, and Acculturation with focus on Reverse Culture Shock. The literature reviewed helps in problem definition and formulation of research hypothesis, by providing detailed insight into the research gap.

Chapter 3: **Sectored Profile and Precursors Effecting Repatriation Adjustment**: This chapter heavily emphasizes the topic of Globalization in the International Human Resource Management in context of Indian corporate sector with expanded emphasis on Repatriate Management and Adjustment in the IT sector. It addresses the broad and cross-disciplinary concept of sectored profile and the precursors effecting Repatriation Adjustment, with focus on the dimension of Indian Repatriate acculturation in multifaceted dimensions in the Information Technology Industry in India. Each construct is introduced and deliberated upon, in context of repatriation of corporate executives, thereby providing firm background to the research hypotheses.

Chapter 4: **Research Design and Methodology**: This chapter deals with the discussion regarding nature and scope of the study, research design, data collection, data purification, reliability and validity, methodology adopted, objectives of the study, and the hypotheses framed. It entails the sample size and selection and the step-by-step procedures used in the research.

Chapter 5: **Data Analysis and Interpretation**: This chapter entails data analysis and interpretation of results derived from applying relevant statistical tools and techniques to the data. The section provides the numerical data and details about the significance of findings. Hypotheses testing and data analysis have been done by the using statistical techniques including factor analysis, multiple regression, and ANOVA. Mean, standard deviation, along with correlation coefficients and reliability test results have been listed in detail in this chapter.

Chapter 6: **Summary and Conclusions**: This chapter highlights the conclusions and suggestions emerging from the discussions. It concludes the present research with a discussion on the meaning of the findings, strategic implications, research contributions, limitations, and recommendations for further research.

Chapter 7: **Policy Recommendations for Better Future**: This chapter Discusses Policy recommendations for a better future.

Contents

About the Authors

Manmeet Bali Nag is currently working as an associate professor, Department of Management and Commerce, School of Management and Commerce, Manav Rachna University, Faridabad. She is an experienced professional with a diverse experience encompassing academics, research, training, organization building, creative writing, and authorship. She has taught MBA, MBE, B.Tech., and Behavioral Sciences scholars, at various institutes, including Nirma University, Ahmadabad. She has been on the editorial team of Wiley group of publishers at Thompson Press.

Her broad areas of teaching include Human Resource Management, Entrepreneurship, Business Communication, Organizational Behavior, and Economics. Besides her academic and journalistic pursuits, she harnesses her skills toward mentoring and coordinating the niche area of "Entrepreneurship, Innovation and Family Business" as an innovation ambassador of IIC MHRD. She has published articles, monographs, chapters, and papers in various national and international journals including Taylor & Francis, Elsevier, Springer, Q1 Category Journals, etc. Her focus areas are human resource dynamics and management, gender and diversity, Repatriation Management, ICT in higher education, emerging trends in business culture, and cross-cultural management. She is credited with an ICSSR Funded Project on elderly care framework. She writes extensively on geopolitics, gender dynamics, and current affairs and has several journalistic articles to her credit. She's is an avid blogger and cultural critic.

Her specialization includes Innovation and Entrepreneurship, Human Resource Management, Economics, Organizational Behavior, Diversity, and Inclusion.

Firdous Ahmad Malik is an Indian citizen by birth, presently working as a senior research associate at Jindal Center for Global South, O. P. Jindal Global University, Sonipat, Haryana. He has graduated Ph.D. from Babasaheb Bhimrao Ambedkar University in Lucknow, India. He holds an M.Phil. degree in Economics from Babasaheb Bhimrao Ambedkar University Lucknow and a Master's degree in Economics from Kashmir University. His research area of interest is Microfinance, financial inclusion, and financial literacy. He has published papers in Springer, Elsevier, Taylor Francis, and other Scopus-indexed journals. He has published his latest book with Springer, *Financial Inclusion Schemes in India.* His second book is published by Notion Press on *Financial Behaviour of Urban Destitues in India.* His third book is published by Booksclinic on *Asymmetry of Information and Lending Risk Livelihood Pattern of Street Vendors in India* and Shineeks published his fourth coming book on *Health System in Jammu and Kashmir Challenges and Opportunities.*

His specialization includes Microfinance, Development Economics, Monetary Economics, Public Finance, and Inclusive Education.

Acronyms

AA	Anticipatory Adjustment
ANOVA	Analysis of variance
BPM	Business Process Management
BPO	Business process outsourcing
CMM	Capability Maturity Model
CSE	Core self-evaluation
CT	Competency Transfer
df	Degree of freedom
EFA	Exploratory factor analysis
ER	Effective Repatriation
GTM	Global talent management
HCN	Home country national
HR	Human Resource
HRD	Human Resource Department
HRM	Human Resource Management
IHRM	International Human Resource Management
IT	Information Technology
IT-BPM	Information Technology—Business Process Management
KMO	Kaiser Meyer Olkin
KSAO's	Knowledge, skills, abilities, and other domains
MNC	Multinational corporation
NASSCOM	National Association of Software and Services Companies
PASW	Professional Analytic Software
PCN	Parent country national
RCS	Reverse Culture Shock
SEZ	Special Economic Zone
Sig.	Level of significance
SPSS	Statistical Package for Social Sciences
STPI	Software Technology Parks of India
TCN	Third country national
U.S	United States

UCT	U—Curve theory
USP	Unique Selling Proposition

List of Figures

List of Tables

Chapter 1
Introduction

1.1 Introduction

The rapid internationalization of organizations in recent years has heightened the interest in cross-cultural paradigms of individuals undertaking international assignments. New challenges have cropped up associated with cross cultural interactions, and in particular the growing need of joint ventures, foreign subsidiary establishment, overseas representative offices and international strategic alliances, has lead to a combination of adaptive management practices. The changing interdependence between IHRM (International Human Resource Management) and other corporate functions (Mendenhall, 2006) has diversified into four key strategic pressures during the Internationalization of HR function which encompasses: maximum shareholder value, forging strategic partnerships, building global presence and creating core business presence. In this context, IHRM has to address the issue of Expatriation and Repatriation on a much wider scale.

For competitive advantage and being critical to success and survival of multinational corporations (MNCs), it is imminent to create and combine the tangible and intangible parameters of the indispensable resource of an MNC; the Human Resources. The exploitation capability of a firm, determines its profitability. Therefore, MNCs send employees from their home country to host country with the objective of position filling, management development, and organization development (Harzing, 2001).

This sets a prelude for global interactions from business point of view, necessitating the inward and outbound movement of personnel. Hence Expatriation and Repatriation and the resultant adjustments, form the core areas of cross cultural study and cross cultural interactions and sojourns. The optimum use of cross cultural teams can enhance the competitive position of organizations (Black & Gregersen, 1991). These parameters include relations between people, motivational orientation, orientation toward risk, definition of self and others, attitudes to time, and attitudes to the environment.

M. B. Nag and F. Ahmad Malik, *Repatriation Management and Competency Transfer in a Culturally Dynamic World*,
https://doi.org/10.1007/978-981-19-7350-5_1

1.2 Evolution of Repatriation Management and Associated Concepts

1.2.1 Expatriation

The process of recruitment in the international context begins with the assignment of tasks to international assignees or Expatriates. The word comes from Latin term Exparitus from *ex* (out of) and *patria* (country or fatherland). A person, temporarily or permanently residing in a country and culture other than that of person's upbringing or legal residence and planning to return to his home country, is called an Expatriate. When a multinational parent company sends a person on a work assignment to a foreign nation, he/she is called an Expatriate (Lee & Liu, 2006). An Expatriate is one who works and lives in a foreign country but remains a citizen of the country where the employing organization is headquartered. The strategic targets addressed by an international assignment or an Expatriate are, to improve business performance by being corporate missionaries, to foster the parent corporate culture in the subsidiary culture, control operations and provide technical and managerial expertise and aid organizational learning, provide developmental strategies and opportunities to self and organization in a cost effective manner, to act as mentors to future Expatriates by imparting skills like transfer of technology and knowledge, managing cross-culturally and negotiating in a specific culture, to break the barriers between parent company and the subsidiaries and development of talent and future leaders of the company by opening up new international markets in a cost effective manner.

As a result of their versatile exposure, Expatriates acquire bi-cultural competence and KSAO's (knowledge, skills, abilities and other domains of expertise) and as such Expatriate talent has manifold benefits as they play a crucial role for their organization and the performance of a company is largely reliant on the performance of its Expatriated and Repatriated employees (Tivey, 2008). Therefore, there is a need of developing managers who understand their global business and have a competitive edge. Inter-subsidiary communication and coordination may be facilitated by international assignments. It also increases the international experience and knowledge base of present and future managers. To gather and maintain a resident base of knowledge, which in turn helps with the complexities of international management, is an important role of an organization.

Post 1991 the Indian government has introduced several reforms and policies to make India a global economic power, which has paved way for various alterations made to open up the system, attract foreign investors, send home country nationals abroad on assignments and to decrease the bureaucratic hassles in implementation of such policies. Inflow and outflow of Expatriates is acknowledged to meet the need of skilled and professional manpower or human expertise in various fields (Harvey, 1982, 1989). This calls for a comprehensive and holistic view of the process of an international assignment in totality. It would harness, optimize and nurture the demographic dividend of India which is an enviable asset that the country possesses.

1.2.2 Repatriate and Repatriation Management

In the era of globalization, organizations have crossed the boundaries of nations in terms of their business operations. Human Resource Management has become dynamic and has paved way for IHRM (International Human Resource Management) as a critical component of strategy for the management of internationally mobile staff which demands new facets of human resource functions to manage in international perspective. Cross-cultural adjustment is the process of adaptation to living and working in a foreign culture. The perceived degree of psychological comfort and familiarity a person has with the new host culture covers the ambit of cross cultural adjustment (Lee & Liu, 2006).

Repatriation of Expatriates remains the twin dimension of an international assignment in the realm of International human resource management. An employee who has returned home after spending more than six months abroad, is called a Repatriate (Linehan and Scullion, 2002). Whilst Repatriation may be considered an integral element of the Expatriation process, it rarely receives the attention that it deserves (Linehan and Scullion, 2002). Organizations are venturing into globally integrated operations which has its share of pressures to adapt and make decisions at local levels (Bartlett & Ghoshal, 1991; Palmisano, 2006), the need to develop a culturally sophisticated workforce and globally savvy business leaders has become pressing. Poor Repatriation is costly (Black et al., 1991), and results in under optimization of the human resource, often leads to loss of talented personnel (Black & Gregersen, 1991), and probably represents the number one reason for employee hesitancy to accept overseas assignments (Harvey, 1989). The result is workforce attrition due to non-management of Repatriates, after the completion of an international assignment. This necessitates cross-cultural preparation of individuals undertaking international assignments, and managing the Reverse Culture Shock and adjustment issues that face Repatriates. It is a new paradigm in international business literature during the past 25 years, and only more recently has Repatriation begun to receive the same serious scrutiny as Expatriation (Gomez-Mejia & Balkin, 1987). There are various factors that can influence the Repatriation of an individual when back in the home country. A valuable source of experience and innovative thinking strategy to enhance the competitive position of organizations is the resultant effect of the effective use of Repatriated employees. The cost of unsuccessful ventures can result in failure of a firm on financial, strategic and other associated parameters of successful Expatriation and repatriation.

Outcome of studies in India which explore the dimensions associated with Repatriation can enhance the prospects of minimizing the problems associated herein and thus maximizing the potential advantages an international assignment offers. From the organizational perspective, Repatriated employees are valuable assets. This is subsequent to the prospective benefits that accrue to an organization, due to proper harnessing of Expatriates and subsequently the Repatriate talent and acumen. Employees bring forth a fresh approach to information gathering, analysis of data, and problem-solving as a result of having worked cross-culturally in an effective manner. They may have become more flexible in changing circumstances and may be able to bring insights and innovation to the planning process. Having performed at a high level in a host culture may bring a dimension of confidence and competence that will enhance the value of an Expatriate, to the company as it competes in a changing world market (Black & Gregersen, 1991). The Repatriated employee has insights due to global exposure that can effect the needed change and may bring a greater tolerance for divergent ideas and more flexible working structure.

The corporate goal is to secure an acceptable return on investment while implementing the company's global strategy. The company's return on investment tends to increase if the company can retain the employees and utilize the experience and knowledge that they have gained. Costs escalate when attrition rates are high and the loss of experienced employees after international assignments results in high training costs for replacements and the loss of return on investment in salaries and benefits paid to the employees while on the assignment (Black & Gregersen, 1991). High cost of assignments and the low returns due to in effective Repatriation, calls for various options for reducing the cost of the international assignments, which includes altering the structure and terms of assignments and revising their Repatriation policy and practice.

The Repatriation process necessarily exhibits itself in four stages of arrival, euphoria, recovery and acculturation and adjustment. Expectancy violation during this process, due to ineffective Repatriation culminates into Repatriation losses and the study deals with an analysis of this phenomenon and its impact on attrition rates and Competency Transfer of human resource. The transfer of global competencies acquired through an overseas assignment is a prerogative of a firm's HR policies and practices. The impact on Competency Transfer of the Repatriate's capacity and desire to adjust to the Repatriation job assignment calls for attention (Caligiuri et al., 2000). HR initiatives for career development and profiling from an overseas assignment, that contributes to the organizational goals of productivity and profit is gaining importance (Bonache et al., 2001). Therefore the role of HR can be summarized in effective selection and pre departure training of Expatriates, effective mentor role and maintenance of communication prior to and after repatriation, provision of necessary help in transition and in overcoming Reverse Culture Shock, providing realistic job preview, keep the perspectives of employee and organization in line, effective career planning and utilization of Repatriate talent and transfer of competency and dual growth of employee and organization.

However, given the intricate and complex dynamics involved herein, companies need to close the gap between expectation and reality facing Repatriates. Emphasis needs to be laid on the intricate fabric of global or cross cultural management and the dynamics which play a pivotal role in its applicability for achieving the laid down objectives of corporate advancement and optimum utilization of resources. In short, companies need to take decisive steps to prevent widespread Repatriate disaffection and it calls for an in depth analysis of the perspective of Repatriation in a corporate set up.

Repatriation Adjustment and Management therefore holds the key to the manifold aspects of utilization of Repatriate talent and competencies. The concept of

Repatriation Adjustment and Management, as a realm of cross cultural management, has evolved on the contemporary global management scene. It refers to the initiatives that capitalize on the re entry of a firm's workforce (including characteristics such as handling Reverse Culture Shock, tapping of talent and career planning, ensure smooth transition from being an Expatriate to a Repatriate, Diaspora option, debriefing sessions for reintegration in home office, family adjustment and re-acculturation, HR initiatives for career development and profiling) from an overseas assignment, as a strategic approach to business that contributes to the organizational goals of productivity and profit. Repatriation adjustment and management refers to the initiatives that capitalize on the re entry of a firm's workforce (Black & Gregersen, 1991).

1.3 The Indian Business Perspective

1.3.1 Globalization and Significance of Repatriation Management in Corporate India

Human Resource Management in the twenty-first Century encompasses the key aspects of the changing landscape of international business and comprises of the prominent issues for IHRM at the beginning of the twenty-first century. The changing nature of the global economic landscape, particularly in India, China and Central and Eastern Europe, the changing nature of careers and, finally, the changing patterns of global staffing, are some of the aspects (Scullion et al., 2007).

As a result of Globalization and liberalization of trade, the political landscape in India also has undergone a sea change in the last two decades. Many MNCs have to interface with the Indian government. MNCs setting up operations in India have found that a lot of emphasis is required for training and motivating employees, doing things in a process-oriented way, and ensuring that the company is customer-oriented. Maximizing the involvement of the parent organization's executives during strategy formulation and project implementation phases has worked well (Krishna & Mitra, 1998). As an employer in India, the company has to be able to demonstrate to its executives and managers that they would be losing something if they left the organization. Motivation is therefore an ongoing process, wherein short term assignments to different parts of the world, provision of job rotation opportunities or job enrichment opportunities, are provided.

Global integration is necessitated for achieving the purpose and this leads to the organization having the right performance focus, management focus, and cultural integration, which is very important for an MNC operating in India, Thus, a focus on operational excellence, innovation, customer excellence, etc., is important for an MNC to gain competitive edge and must be navigated by companies having a global presence. The challenge lies in harnessing the Expatriate and Repatriate talent and preventing its colossal failure that proves to be a bottleneck in employee, organizational and national growth (Gupta & Sivaramakrishnan, 2010).

Several tangibles and intangibles have gone into the making of the success story of Indian Industry. Indian culture has embedded factors and intrinsic aspects which catapulted in the growth of Indian businesses, resulting in the glorious adage—"India shinning". There is a manifestation in full force of the hidden Indian potential once the shackles were removed through the process of liberalization and subsequent paradigms of the Indian economy. Such a heartening scenario has of late ushered in many iconic decisions and implementations which have the forbearance and tenacity of catapulting India on to new horizons. The recent strides in the global business arena are not happenstance. India has been rediscovering itself and racing towards achieving a pre-eminent position that it had once occupied, once the various shackles holding it back, were removed (Dahlman & Utz, 2005).

In Indian perspective, globalization and liberalization makes it imperative for multinational corporations (MNCs) to internationalize their business, wherein they need to offer their employees the possibility of working abroad which is termed as Expatriation. Subsequently Repatriation as a twin facet of an International assignment requires equal and at times more deliberation and thought on the part of the HRD department of an MNC. Several repatriation processes are undertaken by Indian Information Technology (IT) MNCs and attempts should be made to lower the repatriates' turnover intentions (Santosh & Muthiah, 2013). Studies indicate that the surveyed repatriates perceived the support during international assignment and upon return from assignment, as two significant variables to increase their retention, and lack of it thereof was likely to generate unfavorable attitudes toward the company and higher turnover intention (Gupta & Sivaramakrishnan, 2010).

The repatriate's turnover is a matter of great concern if the assignment is used for development of high potential employees, meeting immediate project strategic needs, and/or meeting strategic needs by building a pool of employees for international assignments. Although companies are sending their best employees overseas, they do not manage the repatriation process well, "leaving many to never capitalize on the return on their investment" (Barbian, 2002). International experience fitting into an employee's career progression is not well specified by companies, thereby inadvertently making room for unfounded high expectations. The repatriation process is usually a difficult experience for both the individuals and the organizations (Scullion, 1994). Adjustment problems for both the company and the repatriate, encompasses the Repatriate program. Majority of the repatriates often feel discouraged and angry with the repatriation program resulting in higher turnover rate among expatriates. Therefore it necessitates the need to analyze satisfaction level with the repatriation program in Indian IT sector and to identify its impact on repatriate turnover (Santosh & Muthiah, 2011). Hence the significance of Repatriation Management attains significance in the contemporary scenario.

1.3.2 Growth of the Indian Information-Technology Industry

Post-1991, the entrepreneurial energies of India have been unleashed. The country is experiencing a robust domestic demand helps sustain growth and therefore foreign companies are participating in India's growth. This has been significantly facilitated by tremendous effort on the part of the industry itself to shape up which yielded a tremendous increase in the operational efficiency of the companies and trade with other countries. Over the last few decades India has reached an envious position on the global business front and seems to be coping well with the challenges. While India's contribution to the global trade is still very small, it has grown substantially, of late. The structure of Indian exports has moved from products to services and the share of services in our exports is increasingly growing. Secondly, within the manufacturing space, our growth has been coming from more value-added products than raw materials. Thirdly, the share of knowledge-intensive products going out of India has grown and this marks the onslaught of a dynamic era (Santosh & Muthiah, 2013). Without doubt, post-liberalization, India has gone through a tremendous acceleration of the economy.

Several reforms have been initiated, which have made foreign investment in India, a feasible option. Interestingly, India has achieved a position of pre-eminence, in many spheres of economic activity, most notably in Information Technology, Business Process outsourcing, and various other service industries and also increasingly in manufacturing, since liberalization in the early 90s. Many Indian companies have gone on to take their rightful positions in the world arena and are second to none. Some of them have even emerged as world leaders in their respective domains (Gupta & Sivaramakrishnan, 2010). In India, many multinational companies (MNCs) have head offices overseas, and are increasingly looking at setting up local operations in India to harness the value that the country delivers. Globalization has only aided in spearheading India and success story beyond demarcated borders. According to Roy et al. (2002), rapid advances in information technology (IT) have lead to major changes in the economies of the world. Product quality, Information flexibility and fast response are the key factors for global competition and IT plays a critical role in these areas. In industrialized and developing countries, policy-makers view IT as a critical infrastructure to enhance their access to global knowledge, markets and capital. IT as core capability for development has major implications for infrastructure, transformation of Industry and India becoming a global economy.

The rebounding of India on the global business map had its origins in the Indian IT industry which proved to the world and to other industries in India that it is possible for the country to deliver world class products and services at competitive prices. In terms of the strategic potential of IT and BPO industries, the country still has a long way to go to achieve its full potential. While exports in these sectors are currently about $30 billion, it is estimated that the figure could reach $150 billion by end 2015 which at that time would account for 10–12% of the country's GDP. Today, India is known outside of India as an IT destination and is admired for its well educated human capital. Today, the information technology industry in India has transformed the Brand India. A knowledge revolution in the world and India has lead to the participation of many countries in it because India has the highest knowledge density in the world and it is for companies to harness this knowledge capital and make Indians leaders in the knowledge revolution. In Bangalore, many companies are SEI CMM (Capability Maturity Model), which is a service mark registered with the U.S. Patent and Trademark Office. India has the largest number of companies that are SEI CMM Level 5 certified in the software industry worldwide.

India is the world's largest sourcing destination, accounting for approximately 55% of the US $146 billion market. The country's cost competitiveness in providing Information Technology (IT) services, which is approximately 3–4 times cheaper than the US, continues to be its Unique Selling Proposition (USP) in the global sourcing market. India's highly qualified talent pool of technical graduates is one of the largest in the world and is available at a cost saving of 60–70% to source countries. This large pool of qualified skilled workforce has enabled Indian IT companies to help clients to save US $200 billion in the last five years. India's IT industry amounts to 12.3% of the global market, largely due to exports. Export of IT services accounted for 56.12% of total IT exports (including hardware) from India. The Business Process Management (BPM) segment accounted for 23.46% of total IT exports during financial year 2015. The IT-BPM sector in India grew at a Compound Annual Growth rate (CAGR) of 15% over 2010–15, which is 3–4 times higher than the global IT-BPM spend, and is estimated to expand at a CAGR of 9.5% to US $300 billion by 2020. The Government of India has extended tax holidays to the IT sector for Software Technology Parks of India (STPI) and Special Economic Zones (SEZs). Moreover, the country is providing procedural ease and single window clearance for setting up facilities (www.ibef.org, 2015).

Major players in Indian IT Sector include TCS, which is the largest software company in Asia and was one of the pioneers of the global delivery model. TCS has been seeing strong growth from markets like North America, UK and Europe despite the gloomy economic environment. Infosys is the second largest Indian IT service provider and reputed for its work ethics and world class management practices. By exceeding revenue expectations, it has kept investors and analysts happy. Infosys has an eye for product, consulting, platform space and system integration. Wipro is the third biggest Indian IT firm and its revenues for the year 2011–12 stood at $5.7 billion. It is betting on regions like Latin America, Asia Pacific and Ghana as they have been contributing almost 16–17% of the total revenues for Wipro. HCL Technologies is the fourth largest Indian IT company. In the last 2 quarters, it has won deals worth $2.5 dollars and Europe has contributed almost 54% to it. It is strong in total IT outsourcing which comprises of infrastructure services, application development and maintenance. While most of the other Indian IT firms are vying for high-end consulting contracts, HCL has been focusing on infrastructure deals (Satyamoorthy, 2006). A vast database of IT employees who find themselves as Expatriates in foreign set ups and after a professional stint, return back as Repatriates, going through the cycle of Expatriation and Repatriation is noteworthy in this sector due to the outsourcing of business and also due to the fact that post liberalization India has emerged as a big player on the IT scene (Santosh & Muthiah, 2011).

There is a growing need for global strategic perspectives due to the ever increasing international competition. A global business strategy requires global human resource systems and globally competent people that need to be implemented successfully. Therefore, on a global scale, one of the greatest challenges is the management of human resources. Due to the multi-faceted nature of their employees, human resource personnel need to maximize the potential of each employee. As an attractive method for accumulating foreign markets, organizations perceive global assignments as an important perspective. The greatest concern is the challenges of selection and management of the most appropriate individuals and leveraging their competencies at all stages. Therefore tapping of the Repatriate talent is crucial in terms of the investment associated with it and also due to the fact that competency acquired doesn't go waste. The purview of the research is to analyze the manifold dimensions of Repatriation as it exists in the modern day corporate set up especially in the Information Technology and software industry which has grown leaps and bounds in the recent years.

1.4 Rationale for the Present Study

Expatriation and Repatriation are the two dimensions of an international assignment. Companies need employees that are willing to work globally, since the most effective way to achieve international experience is by living and working within a foreign business arena (Engen, 1995). Due to the outsourcing of business and also due to the fact that post liberalization India has emerged as a big player on the IT scene, a vast database of Indian IT sector employees, find themselves as Expatriates in foreign set ups and after a professional stint, return back as Repatriates, going through a cycle of Expatriation and Repatriation. Indian Expatriates are in increasing demand by Indian enterprises which are growing into regional and global competitors. Effective use of Repatriated employees can provide a source of experience and innovative thinking strategies to enhance the competitive position of organizations. The cost of unsuccessful ventures can result in failure of a firm on financial, strategic and other associated parameters of successful Expatriation and Repatriation. Thus, an overall cycle starting with Expatriation to an analytical Expatriation monitoring and proceeding towards successful repatriation, thus reaping the benefits of the human capital developed in the process, is the consequent prerogative and endeavor of any corporate enterprise.

Therefore tapping of the Repatriate talent is crucial in terms of the investment associated with it and also due to the fact that competency acquired does not go waste. A research oriented approach towards the discussion of Repatriation, is called for, in order to address the issues related to the process, underlining some conditions for successful Expatriation namely providing pre departure training and setting of realistic goals, addressing the culture shock phenomenon, provide training related to the process of reentry, address financial issues, offer a realistic job preview, bring motivations in line with company perspective, provide mentors and career planning along with optimum utilization of Repatriate talent (Black et al., 1999).

The rationale of the present research is to analyze the manifold dimensions of Repatriation as it exists in the modern day corporate set up. Herein, Repatriates have a significant role in organizational learning, as they accelerate the transfer of knowledge from host countries to headquarters, and vice versa. Outcome of studies in India which explore the dimensions associated with Repatriation can enhance the prospects of minimizing the problems associated herein and thus maximizing the potential advantages an international assignment offers. There exists a research gap in the domain of Repatriate Adjustment in India, as there is limited data about the total population of Repatriates in India and its implications on various social and organizational criterion, which if processed and refined further can evolve into something productive and immensely useful in Indian corporate set up. Particularly in the domain of developing nations, the study of Repatriation is still a less researched area. The main focus of this study is to analyze the dimensions of Indian Repatriate acculturation in multi faceted dimensions in the Information Technology Industry in north India. The research analyzes the connect between Reverse Culture Shock and Expectancy violation of adjustment, by analyzing whether Anticipatory Repatriate Adjustment is impacted by Reverse Culture Shock. Competency Transfer, as a significant outcome of Effective Repatriation is also elaborated upon. These intertwined constructs with Indian Repatriates as a backdrop, have not been researched upon earlier and hence the study aims to fill this research gap. A comprehensive study can help to improve the chances of finding the most appropriate actions, meeting the demands of Expatriates, and create a smooth Repatriation process.

1.5 Significance of the Present Study

In the domain of cross cultural interactions reflected in the form of business interpretations, there are numerous studies on Expatriation, but the theme of cross-cultural reentry still remains largely neglected and underestimated in the sojourner's transition trajectory. Available empirical investigations point to a number of crucial concerns, effecting the psychological wellbeing, social readjustment and cultural identity of returning individuals. These concerns strongly suggest that re-entry should become an issue of the highest priority to both sojourning individuals as well as people managing the reentry transitions of travelers.

The main focus of this study is to analyze the dimension of Indian Repatriate acculturation in multi faceted dimensions in the Information Technology Industry in north India. Indian Expatriates are in increasing demand by Indian enterprises which are growing into regional and global competitors. The study also aims at evaluating the Competency Transfer of Repatriates, relating it significantly to Effective Repatriation. Analysis and study of Reverse Culture Shock and anticipatory Repatriate Adjustment, also forms a major component of the study. The proposed topic for the research 'Antecedents of Repatriation: Adjustment and acculturation', aims to fill up the research gap, especially in the context of Indian corporate Repatriates. Particularly in the domain of developing nations the study of Repatriation is still a less researched area.

The significance of this study attains considerable proportions, keeping in view the bigger picture scenario wherein companies are setting up businesses in India, especially in the Software Industry. Therefore the sector must be prepared to make significant investments in training their human resources, both Repatriates and Expatriates, to curb attrition, eliminate wastage of resources, optimize talent, encourage reverse brain drain and harness the global competencies. Thus an overall cycle starting with Expatriation to an analytical Expatriation monitoring and proceeding towards successful repatriation, thus reaping the benefits of the human capital developed in the process, is the consequent prerogative and endeavor this corporate enterprise. A comprehensive study can help to improve the chances of finding the most appropriate actions, meeting the demands of Expatriates, and create a smooth Repatriation process.

1.6 Organization of the Study

The entire study is divided into following six chapters:

This Chapter: Introduction

The first chapter contains an overview of Repatriation Management in the context of a globalized world and from International Human Resource Management (IHRM) perspective, followed by the importance of Repatriate talent and its optimization. It serves as an introduction to the field of Repatriation Management and its antecedents and formalization of what knowledge the research aims to contribute. The chapter delves into the background of Repatriation of executives and conceptualization of Repatriation Adjustment. The chapter follows a discussion explaining the research concept in light of the objectives it aims to achieve, keeping in view the structural formation of the thesis chapters. Repatriation as a strategic Human Resource Management tool for retaining manpower and talent has been highlighted, along with the various nuances which effect and channelize such talent in a particular direction, has been deliberated upon.

Chapter 2: Review of Literature

The second chapter focuses on the review of relevant literature. It contains an expanded discussion on the literature with a complete and comprehensive review of the various researches that have been carried out in the topics of interest for the current study. The purpose of this chapter is to build a theoretical framework for the research by reviewing existing literature. It provides a thorough review of literature related to the Repatriation Adjustment and management, as an intertwined concept relying heavily on associated parameters and domains of Anticipatory Adjustment,

Effective Repatriation and Acculturation with focus on Reverse Culture Shock. The literature reviewed, helps in problem definition and formulation of research hypothesis, by providing detailed insight into the research gap. A detailed account of the variables under study in context of the Repatriation is discussed, which includes reviewing the vast arena of analytical studies associated with it in the realm of IHRM which calls for a deeper probing into the aspect.

Chapter 3: Sectored Profile and Precursors Effecting Repatriation Adjustment

This chapter heavily emphasizes on the topic of Globalization in the International Human Resource Management in context of Indian corporate sector with expanded emphasis on Repatriate Management and Adjustment in the IT sector. It addresses the broad and cross-disciplinary concept of Sectored profile and the precursors effecting Repatriation Adjustment, with focus on the dimension of Indian Repatriate acculturation in multi faceted dimensions in the Information Technology Industry in India. Each construct is introduced and deliberated upon, in context of Repatriation of corporate executives, thereby providing firm background to the research hypotheses.

Chapter 4: Research Design and Methodology

This chapter deals with the discussion regarding nature and scope of the study, research design, data collection, data purification, reliability and validity, methodology adopted, objectives of the study and the hypotheses framed. It details the sample size and selection and the step-by-step procedures used in the research.

Chapter 5: Data Analysis and Interpretation

This chapter entails data analysis and interpretation of results derived from applying relevant statistical tools and techniques to the data. It reports the statistical analyses conducted on the data gathered during the experiment. This section provides the numerical data and details whether the findings were significant. Hypotheses testing and data analysis has been done by the using statistical techniques including Factor analysis, Multiple Regression, and ANOVA. Mean and Standard deviation along with correlation coefficients and reliability test results have been listed in detail in this chapter.

Chapter 6: Summary, Conclusions and Recommendations

This chapter highlights the conclusions and suggestions emerging from the discussions. This chapter concludes the present research with a discussion of the meaning of the findings, strategic implications, research contributions, limitations, and recommendations for further research.

References

Bartlett, C. A., & Ghoshal, S. (1991). What is a global manager? *Harvard Business Review.*

Barbian, J. (2002). Return to sender. *Training, 39*, 40–43.

Black, J. S., & Gregersen, H. B. (1991). When Yankee comes home: Factors related to expatriate & spouse repatriation adjustment. *Journal of International Business Studies, 22*(4), 671–694.
Black, J. S., Gregersen, H. B., Mendenhall, M. E., & Stroh, L. (1999). *Globalizing people through international assignments*. Addison-Wesley Longman.
Black, J. S., Mendenhall, M. E., & Oddou, G. (1991). Toward a comprehensive model of international adjustment: An integration of multiple theoretical perspectives. *Academy of Management Review, 16*(2), 291–317.
Bonache, J. C., Brewster, C., & Suutari. (2001). Expatriation: A developing research agenda. *Thunderbird International Business Review, 43*(1), 3–20.
Caligiuri, P., & Lazarova, M. (2000). Developing effective repatriation policies & practices: The forgotten function in IHRM. *Developing global business leaders*. Greenwood Publishing.
Dahlman, C. J., & Utz, A. (2005). *India & the knowledge economy: Leveraging strengths & opportunities*. books.google.com.
Engen, J. R. (1995). Coming home. *Training, 32*(3), 37–40.
Gomez-Mejia, L., & Balkin, D. B. (1987). The determinants of managerial satisfaction with the expatriation & repatriation process. *Journal of Management, 12*(3), 123–134.
Gupta, A., & Sivaramakrishnan, K.(2010). The state in India after liberalization: interdisciplinary perspectives. - books.google.com
Harvey, M. C. (1982). The other side of foreign assignments: Dealing with the repatriation dilemma. *Columbia Journal of World Business, 17*(1), 53–59.
Harvey, M. G. (1989). Repatriation of corporate executives: An empirical study. *Journal of International Business Studies, 20*(1), 131–144.
Harzing, A. W. (2001). An analysis of the functions of international transfer of managers in MNCs. *Employee Relations, 23*(6), 581–598.
Krishna, P., & Mitra, D. (1998). Trade liberalization, market discipline & productivity growth: New evidence from India. *Journal of Development Economics, 56*(2), 447–462.
Lee, H. W., & Liu, C. H. (2006). Determinants of the adjustment of expatriate managers to foreign countries: An empirical study. *International Journal of Management, 23*(2), 302–311.
Linehan, M., & Scullion, H. (2002). Repatriation of European female corporate executives: An empirical study. *The International Journal of Human Resource Management, 13*(2), 254–267.
Mendenhall, M. E. (2006). Elusive yet critical challenge of developing global leaders. *European Management Journal, 24*(6), 422–429.
Palmisano, S. J. (2006). The globally integrated enterprise, - Foreign Affairs-New York.
Roy, S., Das, T., & Chakraborty, D. (2002). A study on the Indian information sector: An experiment with input-output techniques. *Economic Systems Research, 14*(2), 107–129.
Santosh, B. R., & Muthiah, K. (2011). Repatriation program as a process for retaining repatriates—Issues surrounding international human resource management. *International Journal of Research in Commerce, IT & Management, 1*(3).
Santosh, B., & Muthiah, K. (2013). Turnover intentions among repatriated employees in an emerging economy: The Indian experience. *The International Journal of Human Resource Management, 24*(19).
Satyamoorthy, S. (2006). http://www.mbaskool.com/business-articles/operations/4935-indian-it-sector-an-analysis.html
Scullion, H. (1994). Staffing policies & strategic control in multinationals. *International Studies of Management & Organization, 3*(4), 86–104.
Scullion, H., Collings, D. G., David, G., & Patrick Gunnigle, P. (2007). International human resource management in the 21st century: Emerging themes & contemporary debates. *Human Resource Management Journal, 17*(4), 309–319.
Tivey, T. Laura. (2008). *Reflections on the efficacy of the human resource management strategies of international organizations regarding expatriation, repatriation & relocation* [M.Sc dissertation, University of Nottingham].
www.ibef.org/industry/indian-iT-and-iTeS-industry-analysis-presentation. (2015).

Chapter 2
Conceptual and Theoretical Background

2.1 Introduction

In a globalized world, International assignments have become an integral part of individuals' careers and, for most companies, an indispensable tool for attracting, developing, and retaining talent. There is a large body of research on Repatriation Management and adjustment worldwide with varying nuances, as per the socio-cultural, political and economic perspective of a country. A body of research indicates that an international assignment can be a double edged sword for the individual, as well as the organization. Problems reported in literature include adjustment problems, underperformance, career derailment and losses to the organization due to faulty Expatriation and Repatriation (Black et al., 1999a, 1999b; Kraimer & Wayne, 2004; Tung, 1998). This inability to optimize the cadre of globally trained managers raises questions about the objectives and strategies that drive the investment in Expatriation. Therefore we may define a Repatriate as a person who after spending six or more months abroad, returns home (Linehan & Scullion, 2002). Whilst repatriation may be considered an integral element of the expatriation process, it rarely receives the attention that it deserves (Linehan & Scullion, 2002).

In the Indian context, post liberalization, the corporate sector has grown and diversified leaps and bounds. Indian IT professionals are travelling overseas, to cater to the demands for skilled and professional manpower. The globalization of business has necessitated a holistic analysis of the nuances of international business for organizations. Therefore an in-depth analytical approach supporting the variables effecting repatriation in corporate and its manifold variants is necessitated and compiled herein. The purpose of the literature study is to look at the past work in order to develop meaningful insights into the research area that is under study. This calls for customized research in a comprehensive manner pertaining to Repatriation Management under the derived constructs, as applicable to the Indian IT domain,

M. B. Nag and F. Ahmad Malik, *Repatriation Management and Competency Transfer in a Culturally Dynamic World*,
https://doi.org/10.1007/978-981-19-7350-5_2

as per the reviewed literature, particularly with Black et al. (1992b) framework of Repatriate adjustment as the benchmark and the present study is set out to fill this gap.

2.2 An Overview

A review of literature supporting repatriate management follows a sequence, as repatriation is a consequence of expatriation. In this direction an analysis of research supporting the variables effecting repatriation in corporate and its manifold variants is necessitated and compiled. This chapter discusses the literature covered during the review of the constructs studied in this research. The purpose of the literature study is to look at the past work in order to develop meaningful insights into the research area that is being examined.

Organizations are accelerating towards a globalized world and its associated operations, while catering to the needs and issues at the local level too (Palmisano, 2006). This necessitates the need for global employees and optimization of their potential. Yet, studies consistently show that companies fail to integrate international assignments with long-term career development and succession planning which may result in substantial percentage of Expatriates forsaking the organization upon completion of the international assignment. Multinational corporations (MNCs) invest considerable resources on international assignments, yet many managers report dissatisfaction with their post repatriation careers, and a significant percentage leave the firm within a year. An issue which is crucial from a talent development perspective and has garnered enough research attention in recent years, is the high turnover rate among repatriated international assignees (Lazarova & Cerdin, 2007; Yan et al., 2002).

2.3 Concept of Repatriation and Repatriation Management

There has been a gigantic shift in the essence and dynamics of IHRM (International Human resource management) of late, globally. Over the years, the overseas operations of MNC's have gained significance in generating profits with over 50% of corporate revenues being generated away from the home country (Solomon, 2001). The process of recruitment in the international context begins with the assignment of tasks to International Assignees or Expatriates. As a result of their versatile exposure, expatriates acquire bi-cultural competence and skills in other domains of expertise and as such expatriate talent has manifold benefits due to the important functional role that they play for their organization and the performance of a company is also depends to a great extent on the performance of its expatriated and repatriated employees (Tivey, 2008). Therefore there is a need of developing managers who understand their global business and have a competitive edge. Apart from Expatriation, Repatriation of Expatriates is an essential component of an International assignment.

Research proves that poor repatriation is costly, greatly diminishes the optimization of the human resources (Adler, 1981), often leads to loss of talented personnel (Black & Gregersen, 1991), and probably represents the number one reason for employee hesitancy to accept overseas assignments. The result is workforce attrition due to non-management of repatriates returning to their home country after a global assignment. In international business literature during the past twenty five years, only more recently has repatriation begun to receive the same serious scrutiny as expatriation (Black & Gregersen, 1991; Gomez-Mejia and Balkin, 1987).

The pace at which organizations are accelerating towards globally integrated operations while simultaneously experiencing pressures to adapt and perform at local levels (Bartlett and Ghoshal, 1991; Palmisano, 2006), gives importance to the need for harnessing a globally competent workforce. Repatriation Management therefore holds the key to the manifold aspects of utilization of repatriate talent and competencies. The repatriation and Reverse Culture Shock necessarily exhibits itself in four stages of arrival, euphoria, recovery and acculturation and adjustment. Expectancy violation during this process, due to ineffective Repatriation culminates into repatriation losses and the study deals with an analysis of this phenomenon and its impact on attrition rates and Competency Transfer of human resource.

2.4 Theories Associated with the Concept of Repatriation

In order to explain the interaction of company actions with employee expectations, cultural transitions, and level of commitment to the home organization in the process of expatriation and repatriation, various theories have been propagated by researchers. The theories revealed that employee commitment to the home organization is strongest if they perceive that company management takes care of their career plan that utilizes the experience and knowledge gained during the international assignment, ensures a smooth re-entry transition and maintains continuous communication. This necessitates the need to address the needs of expatriates and their dependents before, during and after their tenure abroad for the successful and efficient integration of the employee into the foreign assignment (Goss & Hynes, 2005).

Basis the findings of these theories, companies have plenty of guidance from researchers and actual best practices to implement a successful global strategy, wherein International assignments become development tools to build the ranks of experienced international managers.

2.4.1 Expectation Theory of Repatriation

Expectation theory emphasizes upon the importance of free flowing communication channels with the employees, regarding jobs they can expect on return to the home office as well as the more common psychological and financial transitions. Martin and Harrell (1996), suggested the expectancy value model which stated that the employees have expectations about their return to their parent organization. Martin and Harrell (1996) advocated that this theory explains much of the employees' dissatisfaction upon return to their home countries. The expectancy violation theory, by Mooradian (2004), states that unfulfilled expectations can be violated either positively or negatively, whereby the levels of repatriation general adjustment, from highest to lowest, occurs, with over met expectations, met expectations, and under-met expectations. Positive evaluation of the repatriation experience, leads to fulfilled expectations whereas unfulfilled expectations lead to negative evaluation and poor readjustment.

The expectancy violation theory (Martin & Harrell, 1996) opined that the unfulfilled expectations can be violated in a positive or negative manner. The employees perceive the experience negatively, if things are negatively inclined than expected, while if things turn out better than expected, they evaluate the experience positively. Therefore, it is better to over prepare the employee and set up realistic (probable or worst case) expectations.

There are numerous suggestions which may strategize the policies in this domain, emanating from the surveys related to this theory: pre-departure training, a realistic job preview, and bringing employee expectations in line with company perspective. This emphasizes the concept of "Expectation Management," which requires the human resources department to cater to the demands of repatriation process.

2.4.2 Re-entry Systems Theory of Repatriation

In continuation with the concept of Repatriate expectations, the Reentry systems theory, as per Martin and Harrell (1996), emphasizes the importance of communication and contact with the employees before, during and after the assignment, wherein the expatriate develops a close association with the headquarters, regardless of geographical distance. Mentors and key individuals, keep the employee updated regarding the changes in organization, personnel, and strategic operations in the home country.

Reentry Systems Theory states that, communication allows the expatriate to feel a close bond with headquarters regardless of geographical distance. Cycle of stress-adaptation-growth through communication, enables the returning employee to adapt to reentry environment. Adler's (1981) study pointed out the importance of communicating organizational changes to employees during the international assignments, by finding out that contact, mentorship and repatriation support aid socialization on return back home.

2.4.3 Reverse Culture Shock and Expectancy Violation Theory of Repatriation

Reverse Culture Shock is a manifestation of the transition period associated with repatriation when "home does not feel like home". The dimensions inbuilt in it are; W curve theory, Degree of social support, Family acclimatization, Identity changes, Cultural shock, Self construal, Expectancy violation (actualization or non actualization of Competing and expected desires, Reward level of competency and communication, Job and lifestyle expectations, Met, unmet and under met expectations on these fronts).

The context herein is an extension of the Expectation theory, where, in addition to hope and distinct from it, repatriates will also possess certain expectations regarding the career prospects that will be awaiting them (Caliguri and di Santo 2001). Based on the theory of met expectations, the more congruent an individual's expectations are with the individual's reality once on the job, the greater the individual's satisfaction and adjustment. Three types of expectations are specified, basis the Expectation theory, namely, under-met, met and over met expectations.

Returning employees often experience Reverse Culture Shock and other difficulties as a result of under-met expectations especially regarding their new job demands. This culminates in a return expectation contrary to what the expatriate might have hoped for. Therefore, such unmet and under-met expectations are concluded as violation of anticipatory repatriate expectations, as a result of the impediments of Reverse Culture Shock.

2.4.4 Equity Theory of Repatriation

Equity theory of repatriation focuses on the employee motivation and commitment dependent on company policy or irrespective which is a deciding factor in determining the retention of the employee. Equity theory ties closely with expectation and reentry systems theories. It relates to employee motivation and commitment to the company, wherein procedural justice refers to the perceived fairness of the process used to determine the distribution of rewards and tends to affect an "employee's organizational commitment, trust in the boss, and intention to quit", as elucidated by Goss and Hynes (2005). This theory helps to explain the degree of commitment that returning employees give to the company. If the employees perceive that the company is not treating them fairly in light of their international experience, they tend to feel less committed.

An examination of factors affecting repatriates' turnover intentions by Lee and Liu (2007) concluded that aligning international assignments with employee career paths reduced the effect of the equity/inequity phenomenon. He predicted repatriate turnover as a test of equity theory. If company management became actively involved in planning for the employee's next assignment upon return, it renders the repatriate's perception of equity as positive, wherein international assignment becomes a stepping stone to career progression. Secondly, he also asserted that management could encourage the positive equity perception by placing the returning employee in a position that utilized the experience and knowledge gained on the assignment. The employees perceived acknowledgement of their experiences and achievements and are more likely to endure the transition from international employee to working within the domestic organization.

2.4.5 U-Curve and W-Curve Theory of Repatriation

W-Curve theory describes the reentry adjustment, often called Reverse Culture Shock as elucidated by Goss and Hynes (2005). The "W curve theory" assimilates the combined U curves; one of Expatriation and the other of Repatriation. The U-curve happens on entry to the country of international assignment. The second U-curve happens when the sojourner returns to the home country. The pattern is the same. The Reverse Culture Shock on return gets aggravated due to the conflicting expectations with respect to the company, community, and cultural changes during his absence. There are also cases when reentry may cause a Reverse Culture Shock if the adjustment phase to the foreign location has been successful, and the return home is not desired. Adler (1981), opined that reentry phase is considerably arduous than the adjustment phase. The associated four stages of adjustment are; the excitement or the honeymoon stage on return and arrival, the hostile and emotional stage, the identity crisis stage and recovery stage and the final adjustment stage (Martin and Harrell, 1996). It is a sum total of the process of initial euphoria, irritation and hostility with cultural differences, adjustment, and reentry to the home country. Facilitation of these stages is a necessity to transfer the acquired competencies of repatriate and acculturate him. In order to retain, harness and accelerate the benefits of repatriation, organizations need to have effective tools of pre departure training, address the Reverse Culture Shock, provide realistic job preview, keep the perspectives of employee and organization in line and strive for effective career planning, thereby facilitating the process of Repatriation.

2.4.6 Hope Theory of Repatriation

As per this theory of career exploration, positive psychological perspective, where career exploration is perceived as productive, can become a determinant of career growth (Shen and Hall, 2009). As a result of a reciprocally-derived sense of successful agency (i.e. goal-directed determination) and pathways (i.e. planning to meet goals), hope is both a disposition and a state that is partially malleable (Snyder, 2000).

The theory also explains the motivational triggers of post assignment career exploration during repatriation. Hope plays a positive role in career exploration and acts as an important determinant of adult career exploration during the repatriation process, whose effect may positively influence repatriate career growth (Zikic et al., 2006). Herein, the role of two important moderators, namely expatriate expectations and social and organizational support for successful repatriation is crucial (Lazarova & Caligiuri, 2001). This transition experience and uncertainty, it carries is likely to trigger repatriate career exploration toward the end of expatriate assignment and upon return to the domestic country (MacDonald & Arthur, 2005). Repatriates with more hope and expectation would be confronted with the disappointment of many unmet expectations.

Thus, expectations and hope, function as a moderating factor, wherein, more the unmet expectations, an individual perceives, especially regarding jobs and career opportunities upon return, the more exploration he conducts in the phase of repatriation. Career growth/success is therefore a valued outcome of repatriate career exploration.

2.4.7 Uncertainty Avoidance/Cultural Dimensions Theory

Uncertainty is a state wherein outcomes and conditions are unknown or unpredictable. Uncertainty avoidance explains the society's tolerance for uncertainty and ambiguity. Some people are more comfortable with uncertainty than others, and the degree to which individuals participate in certain behaviors to stay in comfortable situations is called uncertainty avoidance. Uncertainty avoidance is one of the five cultural dimensions presented by Geert Hofstede in his 1980's book, "Culture's Consequences".

Basis his Cultural dimensions theory, which conceptualized Hofstedes' description of uncertainty avoidance, an index to measure the variable was also developed, by which countries could be compared to one another. Scoring less than expected on the uncertainty avoidance index highlights that people in the country are comparatively comfortable with ambiguity, more entrepreneurial, more likely to take risks, with lesser dependence on structure rules. Countries with high uncertainty avoidance scores desire more stability, more structured rules and social norms, and are less comfortable taking risks. The import of this theory is basis Hofstede's Uncertainty Avoidance Index, wherein the ambiguity related to circumstances may lead to uncertainty avoidance and basis the uncertainty avoidance index, workplace culture can get impacted.

In the cultural paradigm, it indicates the extent to which a culture conditions its members to feel either uncomfortable or comfortable in unstructured situations. Unstructured situations are novel, unknown, surprising, and different from usual. Uncertainty avoiding cultures try to minimize the possibility of such situations by strict laws and rules, safety and security measures, and on the philosophical and religious level by a belief in some absolute Truth.

Four paradigms along which cultural values are ascertained and analyzed, as per the original theory, comprise of: Individualism-Collectivism; Uncertainty avoidance; Power distance (strength of social hierarchy) and Masculinity-Femininity (task orientation versus person-orientation). Hofstede added a fifth dimension, "long-term orientation", to cover aspects of values not discussed in the original paradigm. In the 2010 edition of "Cultures and Organizations: Software of the Mind", Hofstede added a sixth dimension, indulgence versus self-restraint. Further research has refined some of the original dimensions, and introduced the difference between country-level and individual-level data in analysis. Hofstede's work established a major research tradition in cross-cultural psychology and has also been drawn upon by researchers and consultants in many fields relating to international business and communication. It continues to be a major resource in cross-cultural fields.

2.5 Constructs and Concepts Under Study

Keeping in view the flow of the objectives of the proposed study and the four main constructs under consideration, a review the literature is undertaken, sequentially to arrive at outcomes that corroborate the study. This enables to channelize the research towards a desired direction and accelerates the pace of refinement of the study, thereby accessing the gaps in literature which need to be addressed.

The four main constructs chosen viz. Competency Transfer, Effective Repatriation, Anticipatory Adjustment and Reverse Culture Shock are intertwined in literature, so as to arrive at the focal point of Repatriation adjustment.

2.5.1 Acquired Acumen and In-Country Adjustment-Repatriate Competency Transfer

From the organizational perspective, Repatriated employees are valuable assets as they can act as mentors to future expatriates by imparting skills. Organizations that identify a specific career track flowing from a foreign assignment have a more satisfied Repatriate work force, (Naumann, 1993). International assignments are important tools in shaping the perspective and capabilities of dynamic global leaders (Black et al. (1999a, 1999b) providing managers with an opportunity to improve their general management skills, acquire a global mind-set, and build a worldwide network of contacts; the most important variables in a globally inclined organization.

Competitive global economy requires business to have global leaders and an increased internationally skilled workforce to make the best use of all the available resources. Repatriates help to fulfill both the needs as they can help establish and expand global company's international business. They have a highly important role in organizational learning, due to their acumen to transfer knowledge from host countries to headquarters, and vice versa. Companies learn and enhance the knowledge base, when Repatriates share and transfer knowledge and competency (Downes & Thomas, 1999).

A firm's HR policies and practices influence the transfer of global competencies acquired through an overseas assignment (Furuya et al., 2007) which indicates that the competencies acquired through international assignments can be a an important indicator of competitive advantage for firms, and employers need to derive benefits from a repatriate's global competencies by enabling the successful transfer of those competencies through feasible policies and practices, to subsequent job assignments.

Research indicates that the repatriation process is important for a company; bad repatriation leads to dissatisfaction and the risk of former repatriates quitting (Bolino, 2007; Jassawalla & Sashittal, 2009). HR managers need to devise practices to manage this critical resource and therefore, the importance of global sojourns in transferring knowledge and on the associated requirements for HRM practices to ensure the successful and effective retention of expertise as advocated by Bender and Fish (2000). However, companies inadvertently lack expatriate and Repatriation Management practices, thereby failing to integrate international assignments with long-term career development and succession planning (Black et al., 1999a, 1999b).

Companies need to chalk out strategic plans to optimize and assimilate the vast array of repatriates' knowledge and experience, which can be used to enhance competitive advantage in the global marketplace (MacDonald & Arthur, 2005). Theoretical framework of repatriation adjustment, suggests that there are three non-work variables, repatriates need to consider; cultural distance; decrease in social status and housing conditions; and spousal adjustment. Similarly, greater the difference between the home and host cultures, the more the employees have to change. Organizations need to match the level of intensity of their knowledge transfer mechanisms to the type of knowledge gained abroad and promote repatriate motivation to act as agents of knowledge and competence transfer through career development initiatives satisfying repatriate career aspirations (Lazarova & Tarique, 2005).

Expatriates have experienced that their knowledge gained from international assignments is also transferred and exploited by the MNC through informal mechanisms such as networks and own initiatives (Hermansson & Kilnes, 2008). Viable learning of global management competencies on international assignments and their transfer and application in new assignments upon repatriation, happens, via linkages from organizational support, intercultural personality characteristics, self-adjustment, and repatriation policies, leading to heightened job motivation and performance (Furuya et al., 2009). Expatriation and repatriation are important sources of competitive advantage (Velde, 2010), due to the huge amount of knowledge, both tacit and explicit, that corporate may acquire by managing the cycle of reverse knowledge transfer from subsidiaries to headquarter (Canestrino & Magliocca, 2010), and foster corporate learning process (Yuzhe et al., 2011).

Michailova and Mustaffa (2011) demarcated the existing literature into four dimensions, depending on the focus of examination: outcomes of knowledge flows, knowledge characteristics, actors involved in the knowledge flows, and relationships between these actors. Repatriation as a multidimensional phenomenon and the visible benefits of the use of repatriated employees to improve global knowledge and organizational learning in multinational corporations was provided by Kjerfve and McLean (2012).

To create a sustainable competitive advantage, companies need to manage knowledge for competitive advantage and its dissemination to other units and co-workers at the right time and in the right way, a conceptual model from an Indian MNC perspective, to integrate repatriates' knowledge was developed by Santosh and Krishnaveni (2012). Rahman (2013), examined the process of transferring knowledge from a foreign subsidiary to its headquarters, by focusing on subsidiary knowledge creation capabilities.

Minbaeva et al., (2014), discussed the "concept" and the "development" of absorptive capacity of knowledge and acumen, in the host and home country. Emphasis was laid on the effect of the individual differences affecting employees' willingness to engage in knowledge sharing (Caligiuri, 2014). Findings by Wang (2014) inform an interpersonal perspective on knowledge transfer, contributing to work on brokerage, organizational learning, employee mobility, and the globalization of expert knowledge.

Knowledge sharing is the need of the hour and can help companies to create a sustainable competitive advantage, which further necessitates dissemination of knowledge to other units and co-workers at the correct time and in the right way. Whether the Indian MNCs manage to integrate repatriates knowledge, Santosh and Krishnaveni (2015), conducted a study on IT-Repatriates in select South Indian cities, wherein the independent variables of workplace professionalism, employee interaction, source credibility, knowledge quality, source recognition, source awareness, supervisor support, organizational climate, promotion of knowledge exchange, knowledge sharing environment were included. The dependent variable was knowledge sharing environment. A Structural Equation Modeling technique using PLS of empirical data showed that source credibility and organizational climate were the most significant drivers of knowledge sharing environment. Indian IT professionals travel overseas to cater to the demands for skilled and professional manpower, but face adjustment problems, underperformance, career derailment, which may in turn accrue financial losses to the company due to failed expatriation and repatriation. This implies that HR professionals need to chart out tangible career paths, devise efficient performance management systems and challenging job designs so as to motivate employee satisfaction and performance, as an Expatriate and a Repatriate. Therefore an in- depth analytical approach supporting the variables effecting repatriation in IT Sector and its manifold variants is necessitated. Emphasis to develop meaningful insights into the context of repatriate competency and its utilization is required, (Manmeet, 2015). An analysis of the significant relation between Effective Repatriation and transfer of repatriate competency in Indian IT Industry, acquired as a result of overseas assignment was done and the results validated the hypothesis that Competency Transfer is a significant function of Effective Repatriation.

Though MNCs are reasonably good at acquiring knowledge, knowledge as a resource is often wasted through ineffective diffusion and low levels of reuse, resulting in a gap between what the repatriates learn and what their companies as a whole learn. Leveraging the knowledge gained by the repatriates is something that needs further investigation and elaboration. An exploratory and empirical analysis of the dimensions associated with Expatriate acumen acquisition and its consequent transfer as a Repatriate, in Indian IT sector domain, within the ambit of the mentioned criterion, is the aim of the current study. Hence, the present study proposes that Competency Transfer is significantly a function of Effective Repatriation.

Effects of Home Country mentorship on Repatriation Adjustment and Competency Transfer

The effect of home country mentorship on repatriation adjustment has been an important research area in the said domain under study, wherein the repatriates may experience uncertainty and cultural shock (Adler, 1981; Furnham et al., 1988) repatriated from long-distance assignment especially. It was hypothesized that despite the fact that repatriates generally have family and friends in home country, these family and friends may not have living experiences overseas; and thus, are unable to provide suggestion about re-adjustment to repatriates (Mezias & Scandura, 2005), thereby indicating that repatriating back home albeit to a new destination may also limit repatriates' ability to relate feasibly to their previous network of family and friends. Herein, mentors may be the source of solace to the repatriates in overcoming any home country "re-entry shock" (Zeira & Banai, 1987), and a firm-initiated mentoring program may be the only way to ensure that repatriates receive home-country readjustment support (Mezias & Scandura, 2005).

To explain how expatriates internalize these cues, social learning theory by Caligiuri (2000) explains that the repatriates may ask home country mentors about general environmental conditions of home country before their repatriation. Thus, home country mentors may be helpful for repatriates to overcome repatriation adjustment problems. Many a times, repatriates do not associate with the present political landscape and office dynamics (Mezias & Scandura, 2005). Thus, need support, and a mentor can help repatriate reduce pressure of official politics, especially for repatriates of long-distance assignment. According to social learning theory, home country mentors may help repatriates to know official cultures about subsidiary in home country. Thus, home country mentors may be helpful for repatriates to overcome home country official interaction problems.

It conclusively has been corroborated by various research studies that mentoring plays a key role in repatriates' retention and career development (Feldman & Thomas, 1992; Feldman & Tompson, 1993) and facilitates MNCs to consider suitable positions for repatriates due to their increasing human capital as a result of international experiences accumulated by repatriates (Mezias & Scandura, 2005). This can avoid some negative repatriate experiences (Crocitto et al., 2008). According to social learning theory, home country mentors may help repatriates learn work knowledge about headquarters in home country. Thus, it becomes evident that home country mentors may be valuable helpful for repatriates to overcome work adjustment problems.

Crocitto et al. (2008), analyzed the effects of mentors on career outcomes and the impact of home and host country mentors upon expatriate effectiveness. The fact that the host-country mentor has a significant positive effect on the expatriate's organizational knowledge, organizational knowledge-sharing, job performance, promotability, and perceptions of teamwork, was suggested. Post analysis, they opined that mentor from ones' home country has a significant positive effect only on organizational knowledge, job performance, and promotability and significant but negative effect on the expatriate's organization identification and job satisfaction. Neither type of mentoring had a significant effect on job tension.

The effects of mentoring functions on repatriate adjustment–Moderating role of 'Core Self-Evaluations' (CSE) by Wu et al. (2014) analyzed the effects of CSE on home country mentorship and repatriation adjustment. In order to increase self-worth, people who with strong self-esteem will seek assistance of mentor. Pittenger and Heimann (2000) analyzed the self-efficacy of both the mentor and the mentee and advocated the importance of the aspect that it can be a key-determining variable in assessing their readiness and suitability to enter into a mentoring relationship. Turban and Dougherty (1994) argue that people with high self-approval and low negative emotion tend to help others, provide feedback, thus, they keep good mentoring. Erez and Judge (2001) determined that CSE improves the outcomes of task performance as they are positively correlated with goal-setting incentives. To apply this logic to repatriates, in order to re-adjust to headquarters, they view developing home country mentoring actively as an instrument to succeed back to headquarters. Thus, CSE of repatriates has to correlate positively with developing relationship with home country mentors. It has been validated by research that the Repatriates with positive CSE tend to develop stronger home country mentorship. A valuable resource for learning and coping with major organizational changes and value to everyone involved in this relationship is Mentoring. This includes mentees, mentors and the organization. Mentoring functions and their effect on career development and the role of mentoring culture and mentoring structure as moderator, wherein employees working in call centers in India were selected; revealed that mentoring functions significantly affect career development of call centre employees. Further, mentoring culture and mentoring structure moderate the relationship between mentoring functions and career development (Jyoti & Sharma, 2015).

2.5.2 Anticipatory Repatriate Adjustment Expectations and Effective Repatriation-A Perspective

Effective Repatriation is the sum total of all the endeavors, specific to Organizational and home country antecedents, which enable Repatriation adjustment to be worthwhile as per the expected and anticipated standards. Repatriation is defined as the process of individuals' reentry to the home country after living abroad for a significant period of time (Hurn, 1999).

Repatriates experience three different levels of adjustment and General Adjustment, Interaction Adjustment and adjustment to Work (Black et al., 1991). There is a relationship between Anticipatory and In-Country Individual, Job, Organizational and Non-work variables and their relationship with Work, Interaction and General Adjustment (Black & Gregersen, 1991). Vidal et al. (2007) analyzed the key issues affecting the adjustment process of international workers, when back home and the impact of this process on repatriates' performance, satisfaction and turnover intentions when they have returned to their parent companies. The findings confirm that some individual characteristics which comprise of self-efficacy, changes in autonomy at work, creation of working expectations and changes in social status, can affect adjustment to work after the repatriation. Repatriation adjustment process, though a dynamic phenomenon and its effects varying over time, also highlights the importance of reducing the duration of the adjustment process of employees, in order to improve some organizational outcomes. Time variable as a new variable effecting Repatriation process was introduced for model formation.

Theoretical framework of repatriation adjustment, explained through the 15 propositions in Black et al.'s (1992b) theoretical framework of repatriation adjustment provides the key for understanding how repatriation differs from expatriate relocation or a domestic corporate move. Adjustment is categorized into two domains: Anticipatory Adjustments that occur before returning to one's home country; and in-country adjustments that take place after the relocation home. The development of accurate and realistic anticipatory expectations, related to the transition of returning home, will facilitate repatriation adjustment. Anticipatory expectations are related to the notion of preparedness as corroborated by Sussman (2001). Preparedness involves anticipating an event and creating concrete strategies to manage the event. When an event is both predictable and expected, then repatriates have the opportunity for anticipatory preparation and enacting coping strategies, (MacDonald & Arthur, 2005).

This calls for HRD interventions relating to training and development for effective readjustment of international managers on repatriation, Effective Repatriation training programs, training contents, program duration and delivery modes (Ahad et al., 2008). Career planning to career management (Savickas, 2000) which involves goal setting and action planning is required by organizations to meet objectives and by workers to achieve individual career goals. Repatriates who are offered more of these supportive practices, and who believed that such programs were relevant and important for their successful repatriation remain with the company upon repatriation. Considerable percentage of repatriated employees is dissatisfied with the repatriation process (Adler, 1981; Black et al., 1992). Individual related factors also contribute to the in home country adjustment among repatriate spouses and children (Ali, et al., 1995). Many factors influence how an expatriate perceives the repatriation process. The function of international transfer and organizations' beliefs about the value of international experience determining the scope of repatriation programs (Newton et al., 2007).

An Effective Repatriation Strategy (Chew & Debowski, 2008) is necessitated to wedge the gap between intended HR policy and implemented HR practices to reduce the disparity in the academic and practitioner knowledge of the repatriation process of international assignees. Stakeholders include repatriates, Human Resource Business Partners, International Human Resources and line managers (Jassawalla & Sashittal, 2009). HR needs to add transformational practices to facilitate an improved repatriation experience. Selection and management of the most appropriate individuals is the greatest challenge of managing human resources (Singh, 2010). Repatriates reach a level of comfort when all levels of adjustment have been satisfied. Studies revealed several factors associated with Repatriates' improved socio-cultural adjustment (White et al., 2011) which includes role of the extent and nature of support from the employees' own organization.

In Indian MNC's, the success factors of repatriation and their outcomes indicate a strong relationship between intentions to adjust after repatriation with repatriation plans (Santosh & Krishnaveni, 2013). They deliberated upon the repatriation processes undertaken by Indian Information Technology (IT) MNCs and their effectiveness at lowering repatriates' turnover intentions. Basis their analysis, they arrive at the conclusion that adjustments are based on individual initiatives, due to the emotional and psychological tenacity of the expatriates, combined with efforts of peer expatriates, parent firms and host organizations (Chandiok, 2014). The resultant cultural identity changes and the after effects of social support, organizational support and cultural distance between the host and the home country on their international assignment and repatriation experiences, manifests into five central themes: cultural identity independence; knowledge utilization and organizational learning; social network support; global career prospects in the Indian economy; work-life balance was dealt with by Valk et al. (2014).

2.5.3 Repatriation Adjustment and Reverse Culture Shock

Reverse Culture Shock is a documented period of psychological distress that sojourners experience when returning to their country of origin after a long period abroad. Reverse Culture Shock is an area that deserves more attention so sojourners can get the help and understanding that they need to make their reintegration successful. Differences between culture shock and Reverse Culture Shock aren't much, (Gullahorn & Gullahorn, 1963). Only difference between the two is that

Reverse Culture Shock isn't expected. This study provided a platform to suggest future areas of study such as a comparison of cultures and cultural differences that would be beneficial to a greater understanding of the topic in individualist and collectivist cultures with high or low power distance. A primary reason for the return of expatriates before their foreign assignment ends officially, is the uncertainty and frustration resulting from poor cross-cultural adaptation, (Jassawalla et al., 2004).

Focusing on the problems facing an expatriate and the factors leading to repatriation, which included forms of anxiety experienced by an expatriate and his family, certain guidelines were advocated by Murray (1973), for the multinational companies to alleviate the anxieties of the repatriated employees, after an in-depth research of culture shock in reverse experienced by Repatriated executives. This strengthened the belief that Re-entry into an individual's country of origin is more severe than the initial shock they encountered on foreign soil. As early as 1989, Harvey, advocated the necessity of a study directed at the problems of Repatriating employees, programs used by organizations to repatriate those returning and examine why programs have not been instituted by multinational corporations. Expatriate assignments, which many believe will increase career mobility, can actually slow down promotions (Webb, 1996). Return migration and having accurate expectations, in turn, positively affects cross-cultural adjustment (Inkson & Myers, 2003; Watkins, 2013). Moreover, Reverse Culture Shock affects sojourners who are returning from an extended overseas residence and differs between groups and at different life stages (Watkins, 2013). Individuals who adapt and integrate successfully into life in their host country have more severe problems upon re-entry and majority of first-timers experience a degree of Reverse Culture Shock, than those who undertake numerous trips and subsequently have more re-entry transitions leading to learning skills and strategies that lessen the stress of moving between countries (Sussman, 1986).

Alienation, shyness, loneliness and making friends along with general anxiety were all problems that the respondents faced when undergoing Reverse Culture Shock, (Gaw, 2000). Reverse Culture Shock may be an inhibitor for accessing professional help and thus a create hiccups in acculturation and adjustment back home. Repatriation transitions include psychological preparedness, cultural identity, and attributions, (Sussman, 2000) which are further dependent on the preparedness for repatriation, cultural identity change, and attributions of causality on the repatriation experience. It predicted that the repatriation preparedness scale, the psychological subscale of the repatriation experience assessment, and a scale of self-change as an outcome of overseas living are highly internally consistent. Possible identity shifts during transitions happen due to Reverse Culture Shock (Chaban et al., 2011). Acculturation and re-acculturation, happens along with the W-curve and the ten stage model of Reverse Culture Shock, as formulated by Rhinesmith (1975), to assess possible identity shifts during the transition of moving that contribute to identity changes in the participants and how these identity changes do impact on how Reverse Culture Shock affects individuals.

Re-entry leads to expectancy violation (met, unmet and undermet expectations of Repatriation) which is impacted by self-construals, and psychological and social support, (Mooradian, 2004) and assists in understanding and predicting others' behaviours and is applied to Reverse Culture Shock studies to examine how a person's self construal affects their psychological and social adjustment upon re-entry to their home country. The reality of the repatriation process and the expectations of the repatriate are impacted by demographic factors and cultural identity changes (Hyder & Lovblad, 2007) Repatriation experience is dependent on presence or absence of; xenophobic attitudes of colleagues and managers, underutilization of global managers, unplanned transition and career paths. Organizations should reflect on are how they might help expatriates manage the re-entry transition and make a returnees' transition as easy as possible for the repatriate and as beneficial as possible for the company, (Adler & Gunderson, 2008). Return phase of travel is more related to an individual's life course, social relationships and identity rather than their physical return to a geographical location, (Pocock & McIntosh, 2011). Besides, communication is a facilitative tool to eliminate Reverse Culture Shock, (Niesen, 2010) which maybe individual, interpersonal, and cultural. This calls for re-entry orientation and training so that managers don't lose the value of the overseas experiences that the employee is bringing back with them, and lessen the Reverse Culture Shock of the business repatriate.

Socio-Psychological Perspective of Re-entry and Shock

The socio-psychological perspective of re-entry and repatriation of sojourners that underlies beneath the cultural transition, gave insights into the empirical and theoretical literature on cultural transitions, cultural repatriation and the relevant literature on self-concept and identity. Sussman (1986) reported that cultural variations account for patterns in the centrality of cultural identity within one's self-concept, suggesting that some cultures better prepare their members for the experience of being both an expatriate and a repatriate. A framework for studying the individual related factors contributing to the in home country adjustment among repatriate spouses and children (Ali et al, 1995) analyzed the objectives of producing a framework of repatriate spouses and children adjustment. There are many variables that are related to repatriate spouses' and children adjustment to their home country. Individual related factors are variables comprising of personality characteristics, attachment styles, personal preferences and motivation, and personal satisfaction.

Stroh (1997) analyzed the organizational career development predictors of turnover among repatriates, post-return to the parent companies in the past three years. It arrived at the conclusive evidence that attention to career development issues may have an effect on organizations' retention rates, relating to both its domestic and global employees. The variables that are significant predictors of turnover among repatriates are related to the corporate values associated with the importance of an overseas assignment to the organization, whether the organization has a career development plan for repatriates, and the perceived impact of corporate disturbances, on being able to place repatriates adequately upon their return. Hurn (1999), touched upon the physical, psychological and emotional effects of "Reverse Culture Shock", wherein the unrealistic expectations of returnees and their families' called for relevant coping strategies, which includes assistance to the accompanying spouse and children. Changes were suggested in implementation of feasible policies, both from employer and employee perspective.

The relation among self-concept, cultural identity, and cultural transitions was explored by Sussman (2001) who proposed a model that interrelated and analyzed these relations and shifts in cultural identity, which were classified as subtractive, additive, affirmative, or intercultural and it was suggested that all those who participate in cultural transitions are subject to a plethora of experiences collectively labeled as culture shock, adjustment, cross-cultural adaptation, or acculturation. Herein, Culture shock was deciphered as an intense, negative affective response, psychological and physiological, experienced by new expatriates when faced with unfamiliar symbols, roles, relationships, social cognitions, and behavior; Adjustment was defined as the motivational process whereby sojourners attempt to modify both cognitions and behaviors to decrease negatively valenced interactions and experiences and increase the positive ones. Cross-cultural adaptation was defined as the positive consequence of the adjustment process in which cognitions and behavioral modifications produce neutral or positive affect and successful social interactions. Acculturation is the process of long-term adaptation of indigenous groups within plural societies or immigrants to a new culture. It was suggested that Self-Concept and Cultural Identity play a significant role in the above cited parameters and subsequent successful repatriation. Both content and structure theories of "self" reveal that people tend to hold multiple beliefs about themselves based on many self-schemas which include personal attributes about the self (e.g., traits, characteristics, dispositions) and thoughts about membership in social groups such as those formed around gender, ethnicity, social class, religion, and culture. The study has major significance in throwing open several domains of research wherein social psychology is applicable not only to culture as it moves and mixes but to individuals, as cultural beings, as they move and mix between cultures; to the content, evaluation, and structure of the cultural self; and to the awareness of identity in a cultural collective. Sussman (2001) delved into the aspect of the preparedness for repatriation, cultural identity change, and attributions of causality on the repatriation experience. The analysis highlighted the importance of preparedness for repatriation and cultural identity and its impact on repatriation distress. Those repatriates who were the least prepared and had the most cultural identity change experienced more severe repatriate distress. Additionally, confirming that the sojourners attributed the cause of the distress predominantly to situational locus of causality and external control and less to personal control.

Conclusively research arrives at three major observations pertaining to the nuances highlighting Reverse Culture Shock and its consequent implications. The first explores the relationship between psychological preparedness for a repatriation experience and the actual experience itself. Sojourners who had the least preparation for repatriation and therefore presumably the least understanding of what was about to transpire when they returned home, experienced repatriation more distressfully than for those who had a better understanding of the repatriation process. Research focused on changes in cultural identity which predicted that those repatriates who felt less native following an overseas experience would find repatriation more difficult. The third perspective highlights the attribution process of those repatriates for whom repatriation was unexpected and difficult. It was hypothesized that sojourners incorrectly attribute the cause of their repatriation distress as per the situation, under external as compared to personal control, and hence are more unstable. Two major predictors of repatriation distress were preparedness for the experience and cultural identity change.

The relative importance of repatriate support practices offered by organizations and their effectiveness at lowering repatriates' turnover intentions (Lazarova and Caligiuri, 2001), revealed that such measures improved the repatriates' general perceptions about their companies, wherein the Repatriates believed in the relevance and importance of such programs, for their successful repatriation. This strengthened the aspect that their company cared about their overall well-being and ultimately it culminated in a greater desire on their part to remain with the company upon repatriation.

Sussmans model (2001) explored the relationship between cultural identity and repatriation experience. However, the results indicated, unexpectedly, that overseas adaptation and repatriation experiences are not directly in association with each other. Rather, repatriates experiencing high distress reported weak cultural identity because home culture identity strength inversely predicted repatriation distress. It indicated that repatriation experience is related to shifts in cultural identity. As predicted by the Cultural Identity Model, ratings of increased estrangement from parent culture (subtractive) or feeling "more" foreigner (additive) following a sojourn are correlated with the high repatriation distress. Further, the more the global identity shift, the higher the life satisfaction.

Repatriation adjustment research identified potential cross-cultural differences by assessing these processes theoretically and empirically. Focusing on Finnish expatriates and spouses, Gregerson and Stroh (2006), examined anticipatory and in-country variables related to work, interaction, and general repatriation adjustment. The results were that, time overseas, time since returning home, role discretion, and role clarity were significant correlates of repatriates' work adjustment. Time overseas, culture novelty, and social status correlated significantly with repatriates' adjustment to interacting with home country nationals. Time overseas and time back home correlated significantly with repatriate' general adjustment. Time back home, culture novelty, and housing conditions correlated significantly with adjustment to interacting with home country nationals, in case of repatriate spouses. Their general adjustment, time overseas, time back home, culture novelty, and housing conditions were significant correlates.

Impact of Culture Shock on Individual Psychology was analyzed by Xia, (2009), who took into account the negative effects of culture shock into account and paid attention to minimizing psychological discomfort when entering new cultural patterns. It describes a number of solutions and evaluates the effectiveness of them. Considering that character and disposition may be various among different people, it indicated that in order to attain satisfactory results, choosing appropriate methods and reducing psychological stress to the controllable level are very important. The purpose was to describe the impact of culture shock on individual psychology and give specific solutions by citing the main reason for culture shock, the negative impact of culture shock on individual and the corresponding solutions from two stages: pre-departure and during the experience. The first stage emphasizes how to make preparations for culture shock before going abroad, and the second stage stresses on how to overcome psychological stress when people experience culture shock.

2.5.4 *Repatriation Management and the "HR Perspective"*

HRM is a set of policies that are designed to maximize organizational integrations, employee commitment, flexibility and quality of work, (Guest, 1987). Such a policy would call for an expatriate policy that outlines the companys' intention towards the individual and more specifically which is inclusive of Repatriation and the tasks undertaken by the company and of the Repatriate. It was argued that HRM implies giving personnel away to line management (Guest, 1987) which would eventually create a dilemma for HR practitioners keen on promoting HR at board level but find it difficult to implement the same, as this involves commitment from line management. Thus any policy or practice would require the support of the line manager. Within international assignments it is generally accepted (Black et al., 1992) that all expatriates will undertake an assignment that will be administered under a particular policy such as a short term assignment, long term business trip etc. The policy consists of information related to the assignment such as: relocation and costs and benefits; allowances; educational assistance; resignation or dismissal; temporary accommodation; redundancy and redeployment etc. However, the actual practices such as repatriation allowances, assistance with housing, education, shipments etc. are perceived by the expatriate and to what extent there is flexibility towards practices adopted during repatriation is less clear, as suggested by literature. It may be evident enough that family support can be more prolific for a returning. Therefore, HR practices can provoke failure and cause assignees to return from assignment early. This implies that the policies and practices as perceived by Repatriates can form an interesting area for further research including the nuances related to the paradigm, as to why some people are more inclined to undertake an assignment in the first place, which may affect the way they view and ultimately experience the repatriation process, which involves a plethora of associated factors.

Boselie et al. (2005) furthered the discussion within the literature regarding HR practices and policies, to arrive at the definition where best practice refers to a set of HR policies that arguably improves performance in all organizations. In contrast, the best fit approach argues that if HR policies are consistently adopted with the business strategy then the performance improves. Kinnie et al. (2005) found that each group demonstrated preferences for differing HR practices, by studying three groups: managers, professionals and workers. The commitment of managers was associated with four HR practices, with the most prevalent one being linked to career opportunities with rewards and recognition, involvement, communication while work life balance was secondary in importance. On the other hand, professionals' main emphasis was on performance appraisal, though they also valued these practices. Workers, on the other hand, valued rewards and recognition. The conclusion from the said study implied that, different HR practices are more relevant to different employee groups, thus, affecting their commitment to the organization. As a consequence, an employees' satisfaction is linked to different HR practices, which may impose challenges for HR strategy to deliver a solution that fits diverse requirements. The covering argument, to maintain a consistent HR approach, though mandatory with regard to legal implications as well as from an ethical view, becomes less clear with regard to the particular approach that would work for all HR practices and thus be uniform in applicability.

As a result of the onslaught of international assignments, further complexity was noted by Paik et al. (2002) with respect to HR practices applicable to Repatriates. The research found that whereas the Repatriates placed greater emphasis upon reintegration with the home country culture, the HR managers in contrast placed high emphasis upon reintegration with the home corporate culture. This necessitated the focus on communication with the home location, as a crucial aspect of the cyclical process of an International assignment. This would avoid a problematic reintegration with the home location (Paik et al., 2002). Discrepancies often exist between the motivations and expectations of HR managers and assignees suggesting that employee and employer expectation may or may not be aligned. Furthering the argument, Hailey et al. (2005) stated that the HR department needs to go beyond designing effective HRM policies and practices to ensure that these practices are implemented appropriately and are accepted by employees in order to achieve the intended results. In totality, the HR department and line management together have a crucial role to play in stimulating appropriate employee behavior on behalf of the firm (Bossard et al., 2005; Lazarova & Cerdin, 2007).

In the case of Repatriates, research into the role of the HR department offers further complexity. The HR administration of an international assignment requires additional HR practices to support the Expatriate and Repatriate (Williams and Shaw, 2005). introduced the idea that employees respond attitudinally and behaviorally to HR practices, indicating that the employees' perceptions of and reactions to HR practices vary considerably across the organization and therefore the purpose of the practice may not achieve the intended result (Providing pre-departure cross cultural training may not be valued where the assignee does not see the benefit. This view is supported by a recent review that suggests the perceptions of HRM professionals, line managers and employees may differ, but also because of their differing interests. Kinnie et al. (2005) suggested that research should seek to explore the links between actual HR practices and policies as perceived from an individual and organizational perspective. It was indicated that for different employee groups, different HR practices are important for stimulating affective commitment to the organization. With regard to expatriate commitment, Furuya et al. (2007) noted that the policies and practices exerted as part of the Repatriation process only add to the perception of an assignee that the organization values their contribution and thus enhance Expatriates' commitment towards the organization. It is vital to look into HR policies and practices when one seeks to explain Repatriate commitment whilst acknowledging that the policies and practices are only as good as they are perceived. Black et al. (1992) highlighted three problematic areas for the international assignee and their families with regard to Repatriation Adjustment, viz; finding a new position, communicating with home country colleagues and friends and the general culture of the home country. This revelation confirms the role and need of providing supporting practices to identify a return position upon return, keep the communication channels open to maintain contacts with family and colleagues as well as link to the prospective return position and provide assistance to facilitate home country visits during the assignment. This forms an essential element in reducing uncertainty and increasing receptivity towards an assignment. Most importantly, attention to Reverse Culture Shock and organizational intervention to limit the possible effects of culture in the home country was paid by (Baruch & Altman, 2002; Black & Gregersen, 1991). Therefore identifying better supporting practices helps to develop more grounded expectations of Repatriation (Allen & Alvarez, 1998; Suutari & Brewster, 2003).

Considerable research into ongoing support activities has focused upon the support offered by coaching and mentoring relationships (Black et al., 1992; Brewster et al., 2007; Lazarova & Caligiuri, 2001). Moreover, Suutari and Valimaa (2002) argue that Repatriation consists of different factors influencing general adjustments that include the expatriates' age, length of assignment, timing of the negotiations, expatriation adjustment problems, and contact with home and role conflict. Thus the particular supporting practices offered by the organization and knowledge of the repatriates' demographics may impact the support required. Nonetheless, differing expectations between the assignee and the organization also impact upon the Repatriation, extending to non-work issues such as job assistance for the returning partner (Paik et al. 2002). Differing expectations between the organization and the individual can influence the repatriates' commitment to the organization he/she returns to. This necessitates that the expectations of HR supportive practices with the Repatriates' expectation, need to be matched, further impacted by how the practices and policy are delivered, whilst the employee is on an International Assignment. The role of the corporate human resource (HR) function in multinational corporations regarding global talent management (GTM) remains a less researched area. GTM is explored here from two perspectives: increasing global competition for talent, which considers the mechanisms of GTM, and new forms of international mobility, which emphasizes on individual willingness to be mobile, especially in emerging markets, and the organizational capability needed to manage this talent. New corporate HR roles need to be identified to address such issues, so that research may advance the understanding of GTM theory and practice by considering the major future challenges facing corporate HR (Scullion et al. 2010).

2.6 Research Gap Analysis

Indian software and services' exports have been rising rapidly over the last decade, thereby leading to gains from trade. The growth in this sector is driven by steady increase in scale and depth of existing service lines; addition of newer vertical-specific and emerging niche business services, continued expansion of service portfolios and higher value processes. IT professionals form a large chunk of the total knowledge workers working globally and their motivations are largely guided by their perceptions of organizational support, both by intrinsically challenging jobs and a certain degree of self management on the job. This in turn implies that HR professionals need to chart out tangible career paths, devise efficient performance management systems and challenging job designs so as to motivate employee satisfaction and performance, as an Expatriate and a Repatriate (Webb, 1996; Morgan et al. 2004).

Our findings take into account the Repatriates' perspectives in anticipation of expectations and reality, as manifested in Reverse Culture Shock, along with Competency Transfer and sharing, basis the paradigm of Effective Repatriation and Repatriate adjustment. The data was collected from IT sector repatriates via a Questionnaire. The study is imperative of the fact that Indian MNCs use international assignments in order to positively respond to the increasing internationalization in the economic areas in which they operate. Evidence reveals that expatriation is seen as a strategy to fulfill main organizational objectives through their expatriates (e.g., business internationalization, improvement of the coordination and control level of the units/subsidiaries abroad, replication of aspects of the home base, development and incorporation of new organizational techniques and processes). Indian MNCs have developed an International Human Resources Management strategy, wherein Expatriates and Repatriates have a central role in transmitting culture and technical knowledge from company's headquarters to the company's branches and vice versa.

The international presence of this workforce makes it pertinent for multinational corporations to compete globally and build a pool of international management expertise and incorporate appropriate features to their management of repatriates and thereby reap the benefits of decreasing employee turnover through their success in the global arena. This calls for a holistic analysis of the corporate repatriation scenario in Indian context after an in-depth analysis of the relevant literature as has been condensed above to give an overview of the research done and the consequent narrowing down of the study, as relevant in the study of Repatriation adjustment and its antecedents, as applicable to the Indian IT sector.

Literature pertaining to Repatriation Management indicated that there are several questions that need to be answered. Given the context of Repatriation theories of Uncertainty avoidance, Reverse Culture Shock and Expectancy violation, Expectation, Equity and Hope theory, the constructs of the present research delve deeper into the precincts of the variables related to these theories. It analyzes the connect between Reverse Culture Shock and Expectancy violation of adjustment, by researching whether Anticipatory Repatriate Adjustment is impacted by Reverse Culture Shock which disenables meeting of expectations, in the Indian corporate context. Similarly Equity, Expectation and Hope theory has been corroborated to the study with regard to Competency Transfer antecedent, as a significant outcome of Effective Repatriation. These twin areas form a viable realm of study, with Indian Repatriates as a backdrop, not researched upon earlier. There exists a research gap in both the hypothesized realms, in the Indian IT sector Repatriation context and the present study aims to fill it. Therefore it necessitates the study, keeping in view the socio-political and cultural nuances, peculiar to the Indian Corporate milieu and the recent spate of Repatriation back to India in the event of the countrys' economic growth and impetus to "Make in India". It necessitates a thorough research on the optimization of such talent, who otherwise can fall prey to reverse brain drain, thereby justifying the findings.

This calls for a comprehensive HR strategic initiative after a thorough analysis of the variables and constructs at play. The study aims at evaluating the Competency Transfer of Repatriates in the Indian Information Technology Sector, by relating it significantly to Effective Repatriation analysis. Study of Reverse Culture Shock and Anticipatory Repatriate Adjustment, constitutes the other major construct of the study. Subsequent to an in-depth study of literature and statistical analysis of the constructs undertaken, an Industry specific strategic model of Effective Repatriation is consequently proposed.

References

Adler, N. J. (1981). Re-entry: Managing cross-cultural transitions. *Group & Organizational Studies, 6*(3), 341–356.

Adler, N. J., & Gunderson, A. (2008). *International dimensions of organizational behavior*. Thomson South Western.

Ahad, M., Osman, G., & Akmal, S. H. (2008). Repatriation readjustment of international managers: An empirical analysis of HRD interventions. *Career Development International, 13*(5), 456–475.

Ali, J. A., Hin, W. C., & Yusoff, Y. (1995). A framework for studying the individual related factors contributed to the in home country adjustment among repatriate spouses & children. Universite Sains Malaysia.

Allen, D., & Alvarez, S. (1998). Empowering expatriates & organizations to improve repatriation effectiveness. *Human Resource Planning, 21*(4), 29–39.

Bartlett, C. A., & Ghoshal, S. (1991). What is a global manager? *Harvard Business Review*.

Baruch, Y., & Altman, Y. (2002). Expatriation & repatriation in MNCs: A taxonomy.

Bender, S., & Fish, A. (2000). The transfer of knowledge & the retention of expertise: The continuing need for global assignments. *Journal of Knowledge Management, 4*(2), 125–137.

Black, J. S., & Gregersen, H. B. (1991). When Yankee comes home: Factors related to expatriate & spouse repatriation adjustment. *Journal of International Business Studies, 22*(4), 671–694.

Black, J. S., Gregersen, H. B., & Mendenhall, M. E. (1992). Toward a theoretical framework of repatriation adjustment. *Journal of International Business Studies, 23*(4), 737–760.

Black, J. S., Gregersen, H. B., Mendenhall, M. E., & Stroh, L. (1999a). *Globalizing people through international assignments*. Addison-Wesley Longman.

Black, J. S., Gregersen, H. B., & Morrison, A. (1999b). *Global explorers: The next generation of leaders*. Routledge.

Black, J. S., Mendenhall, M. E., & Oddou, G. (1991). Toward a comprehensive model of international adjustment: An integration of multiple theoretical perspectives. *Academy of Management Review, 16*(2), 291–317.

Bolino, M. (2007). Expatriate assignments & intra-organizational career success: Implications for individuals & organizations. *Journal of International Business Studies, 38*, 819–835.

Boselie, P., Dietz, G., & Boon, C. (2005). Commonalities and contradictions in HRM & performance research. *Human Resource Management Journal, 15*(3), 67–94.

Bossard, A. B., Richard, B., & Peterson, R. B. (2005). The repatriate experience as seen by American expatriates. *Journal of World Business, 40*(1), 9–28.

Brewster, C., Sparrow, P., Vernon, G., & Houldsworth, E. (2007). International human resource management. cipd.co.uk.

Caligiuri, P. (2014). Many moving parts: Factors influencing the effectiveness of HRM practices designed to improve knowledge transfer within MNCs. *Journal of International Business Studies, 45*, 63–72.

Caligiuri, P., & Di Santo, V. (2001). Global competence: What is it, & can it be developed through global assignments? *Human Resource Planning, 24*(3), 27–36.

Canestrino, R., & Magliocca P. (2010). Managing expatriation, repatriation & organisational learning in MNC'S: An integrative framework. *Review of International Comparative Management, 11*(2).

Chaban, N., Williams, A., Holland, M., Boyce, V., & Warner, F. (2011). Crossing cultures: Analyzing the experiences of NZ returnees from the EU (UK vs. non-UK). *Journal of Intercultural relations, 35*(6), 776–790.

Chandiok, S. (2014). Perception of IT professionals about cross cultural environment—A study on Indian IT professionals of Delhi & NCR. *European Journal of Commerce*.

Chew, J., & Debowski, S. (2008). Developing an effective repatriation strategy for MNC: A model & tools for international human resource management. *Journal of Comparative International Management, 11*(2).

Crocitto, M., Carraher, S., & Sullivan, S. (2008). Mentoring across global boundaries: An empirical examination of home & host country mentors on expatriate career outcomes. *Journal of International Business Studies, 39*, 1310–1326.

Downes, M., & Thomas, A. (1999). Managing overseas assignments to build organizational knowledge. *Human Resource Planning, 22*(4), 33–48.

Erez, A., & Judge, T. A. (2001). Relationship of core self-evaluations to goal setting, motivation, & performance. *Journal of Applied Psychology, 86*(6), 1270–1279.

Feldman, D., & Thomas, D. (1992). Career management issues facing expatriates. *Journal of International Business Studies, 23*(2), 271–293.

Feldman, D., & Tompson, H. (1993). Expatriation, repatriation, & domestic geographical relocation: An empirical investigation of adjustment to new job assignments. *Journal of International Business Studies, 24*(3), 507–529.

Furnham, A., Kim, A., Yun, Y., & William, B. (1988). The adjustment of sojourners. Cross-cultural adaptation: Current approaches. *International & Intercultural Communication Annual, 11*, 42–61.

Furuya, N., Stevens, M. J., Bird, A., Oddou, G., & Mendenhall, M. (2009). Managing the learning & transfer of global management competence: Antecedents & outcomes of Japanese repatriation effectiveness. *Journal of International Business Studies, 40*, 200–215.

Furuya, N., Stevens, M. J., Oddou, G., Bird, A., & Mendenhall, M. E. (2007). The effects of HR policies & repatriate self-adjustment on global competency transfer. *Asia Pacific Journal of Human Resources, 45*(1), 6–23.

Gaw, K. F. (2000). Reverse culture shock in students returning from overseas. *International Journal of Intercultural Relations, 24*(1), 83–104.

Gomez-Mejia, L., & Balkin, D. B. (1987). The determinants of managerial satisfaction with the expatriation & repatriation process, *Journal of Management*.

Goss, Y., & Hynes, G. E. (2005). Strategies for successful repatriation. *Journal of International Business Research, 4*(1), 97–108.

Guest, D. E. (1987). Human resource management & industrial relations. *Journal of Management Studies, 24*(5), 503–521.

Gullahorn, J. T., & Gullahorn, J. E. (1963). An extension of the U-curve hypothesis. *Journal of Social Issues, 19*(3), 33–47.

Hailey, H. V., Farndale, E., & Truss, C. (2005). The HR department's role in organizational performance. *Human Resource Management Journal, 15*(3), 49–66.

Hermansson, F., & Kilnes, U. (2008). Knowledge transfer from expatriates: A study of MNCs' exploitation of expatriates' knowledge. Uppsala University, Faculty of Social Sciences, Department of Business Studies.

Hurn, B. J. (1999). Repatriation—The toughest assignment of all. *Industrial & Commercial Training, 31*(6), 224–228.

Hyder, A., & Lovblad, M. (2007). The repatriation process—A realistic approach. *Career Development International, 12*(3), 264–281.

Inkson, K., & Myers, B. (2003). The big OE-Self directed travel & career development. *Career Development International, 8*(4), 170–181.

Jassawalla, A., Connelly, T., & Slojkowski, L. (2004). Issues of effective repatriation: A model & managerial implications. *S.A.M. Advanced Management Journal, 69*(2), 38–46.

Jassawalla, A. R., & Sashittal, H. C. (2009). Thinking strategically about integrating repatriated managers in MNCs. *Human Resource Management, 48*(5), 769–792.

Jyoti, J., & Sharma, P. (2015). Impact of mentoring functions on career development: moderating role of mentoring culture & mentoring structure. *Global Business Review, 16*(4), 700–718.

Kinnie, N., Hutchinson, S., Purcell, J., Rayton, B., & Swart, J. (2005). Satisfaction with HR practices & commitment to the organization: Why one size does not fit all. *Resource Management Journal, 15*(4), 9–29.

Kjerfve, T., & McLean, N. G. (2012). Repatriation of expatriate employees, knowledge transfer, & organizational learning: What do we know? *European Journal of Training & Development, 36*(6), 614–629.

Kraimer, M. L., & Wayne, S. J. (2004). An examination of perceived organizational support as a multidimensional construct in the context of an expatriate assignment. *Journal of Management., 7*(3), 213–226.

Lazarova, M., & Caligiuri, P. M. (2001). Retaining repatriates: The role of organizational support practices. *Journal of World Business, 36*(4), 389–401.

Lazarova, M., & Cerdin, J. (2007). Revisiting repatriation concerns: Organizational support versus career & contextual influences. *Journal of International Business Studies, 38*(3), 404–429.

Lazarova, M., & Tarique, I. (2005). Knowledge transfer upon Repatriation. *Journal of World business, 40*(4), 361–373.

Lee, H. W., & Liu, C. H. (2007). An examination of factors affecting repatriates' turnover intentions. *International Journal of Manpower, 28*(2), 122–134.

Linehan, M., & Scullion, H. (2002). Repatriation of European female corporate executives: An empirical study. *The International Journal of Human Resource Management, 13*(2).

Macdonald, S., & Arthur, N. (2005). Connecting career management to repatriate adjustment. *Career Development International, 10*(2), 145–158.

Manmeet, B. N. (2015). Is repatriate competency transfer, a consequence of effective repatriation? An analysis of Indian IT-sector repatriates: The paradigms at play. *Sai Om International Journal of Commerce & Management, 2*(7), 46–55.

Martin, J. N., & Harrell, T. (1996). Reentry training for intercultural sojourners. In *Handbook of intercultural training* (pp. 307–326). Sage.

Mezias, M. J., & Scandura, T. A. (2005). A needs-driven approach to expatriate adjustment & career development: A multiple mentoring perspective. *Journal of International Business Studies, 36*, 519–538.

Michailova, S., & Mustaffa, Z. (2011). Subsidiary knowledge flows in multinational corporations: Research accomplishments, gaps, & opportunities. Department of Management & International Business. The University of Auckland, Business School, New Zealand.

Minbaeva, D. B., Pedersen, T., Bjorkman, I., & Fey, F. C. (2014). A retrospective on: MNC knowledge transfer, subsidiary absorptive capacity, & HRM. *Journal of International Business Studies, 45*, 52–62.

Mooradian, B. L. (2004). Going home when home does not feel like home: Reentry, expectancy violation theory, self construal and psychological & social support. *Intercultural Communication Studies., 24*(3), 123–140.

Morgan, L., Nie, W., & Scott, T. Y. (2004). Operational factors as determinants of expatriate & repatriate success. *International Journal of Operations & Production Management, 24*(12), 1247–1268.

Murray, J. A. (1973). Repatriated executives: Culture shock in reverse. *Management Review, 62*(11), 43.

Naumann, E. (1993). Antecedents and consequences of satisfaction & commitment among expatriate managers. *Group & Organization Management.*

Newton, S., Hutchings, K., & Kabanoff, B. (2007). Repatriation in Australian organizations: Effects of function & value of international assignment on program scope. *Asia Pacific Journal of Human Resources, 45*(3), 295–331.

Niesen, C. C. (2010). Navigating reentry shock: The use of communication as a facilitative tool. University of New Mexico, Department of Communication & Journalism.

Paik, Y., Segaud, B., & Malinowski, C. (2002). How to improve repatriation management: Are motivations & expectations congruent between the company & expatriates? *International Journal of Manpower, 23*(7), 635–648.

Palmisano, S. J. (2006). The globally integrated enterprise, Foreign Affairs-New York.

Pittenger, K. K. S., & Heimann, B. A. (2000). Building effective mentoring relationships. *Review of Business*, 38–42.

Pocock, N., & McIntosh, A. (2011). The return from travel: a new beginning. *Current Issues in Tourism, 14*(7).

Rahman, M. (2013). *A case study investigation of a model for effective reverse knowledge transfer* [Thesis, Master of Business, Auckland University of Technology].

Rhinesmith, S. H. (1975). Bring home the world: A management guide for community leaders of international programs. library.wur.nl.

Santosh, B. R., & Krishnaveni, M. (2012). Knowledge transfer from repatriated employees: The Indian experience. *IUP Journal of Knowledge Management, 10*(1), 7–25.

Santosh, B. R., & Krishnaveni, M. (2013). Turnover intentions among repatriated employees in an emerging economy: The Indian experience. *International Journal of Human Resource Management., 24*(19), 3667–3680.

Santosh, B. R., & Krishnaveni, M. (2015). Knowledge sharing by the repatriates: Experience from Indian information technology (IT) sector. *Journal of Contemporary Research in Management, 10*(1), 31–44.

Savickas, M. L. (2000). Renovating the psychology of careers for the twenty-first century. In *The future of career*. books.google.com.

Scullion, H., Sparrow, P., & Farndale, E. (2010). The role of the corporate HR function in global talent management. *Journal of World Business, 45*(2), 161–168.

Shen, Y., & Hall., D. T. (2009). When expatriates explore other options: Retaining talent through greater job embeddedness & repatriation adjustment. *Human Resource Management, 48*(5), 793–816.

Singh, S. (2010). The journey of expatriates from pre expatriation to post expatriation. *Review of International Comparative Management, 11*(2).

Snyder, C. R. (2000). The past & possible futures of hope. *Journal of Social & Clinical Psychology.*

Solomon, E. (2001). The dynamics of corporate change: Management's evaluation of stakeholder characteristics. *Human Systems Management, 20*(3), 257–265.

Sussman, N. M. (1986). Re-entry research & training: Methods & implications. *International Journal of Intercultural Relations., 10*(2), 235–254.

Sussman, N. M. (2000). The dynamic nature of cultural identity throughout cultural transitions: Why home is not so sweet. *Personnel & Social Psychology Review, 4*(4), 335–373.
Sussman, N. M. (2001). Repatriation transitions: Psychological preparedness, cultural identity, & attributions among American managers. *International Journal of Intercultural Relations, 25*(2), 109–123.
Suutari, V., & Brewster, C. (2003). Repatriation: Empirical evidence from a longitudinal study of careers & expectations among Finnish expatriates. *International Journal of Human Resource Management, 14*(7), 1132–1151.
Suutari, V., & Valimaa, K. (2002). Antecedents of repatriation adjustment: New evidence from Finnish repatriates. *International Journal of Manpower, 23*(7), 617–634.
Tivey, T. L. (2008). *Reflections on the efficacy of the human resource management strategies of international organizations regarding expatriation, repatriation & relocation* [M.Sc dissertation]. University of Nottingham.
Tung, R. L. (1998). American expatriates abroad: From neophytes to cosmopolitans. *Journal of World Business.*
Turban, D. B., & Dougherty, T. W. (1994). Role of protégé personality in receipt of mentoring & career success. *Academy of Management Journal, 37*(3), 688–702.
Valk, R., Velde, M., Engen, M., & Godbole, R. (2014). International career motives, repatriation & career success of Indian women in science & technology. *Journal of Global Mobility, 2*(2), 203–233.
Velde, C. R. (2010). Intercultural knowledge management: exploring models for repatriation competency in the global workplace. *International Journal of Human Resource Development & Management, 10*(4).
Vidal, M., Valle, R., Aragón, M. I. B., & Brewster, C. (2007). Repatriation adjustment process of business employees: Evidence from Spanish workers. *International Journal of Intercultural Relations, 31*(3), 317–337.
Wang, D. (2014). Activating cross-border brokerage: Inter-organizational knowledge transfer through skilled return migration. *Administrative Science Quarterly.*
Watkins, T. L. (2013). *Welcome home? New Zealanders experiences of return migration.* Massey University, Albany, New Zealand.
Webb, A. (1996). The expatriate experience: Implications for career success. *Career Development International, 1*(5), 38–44.
White, D., Absher, K., & Huggins, K. (2011). The effects of hardiness & cultural distance on sociocultural adaptation in an expatriate sales manager population. *Journal of Personal Selling & Sales Management, 31*(3), 325–337.
Williams, A., & Shaw, G. (2005). Knowledge transfer & management in tourism organizations: An emerging research agenda. *Tourism Management, 30*(3), 325–335.
Wu, M., Zhuang, W. L., & Hung, C. (2014). The effects of mentoring functions on repatriate adjustment–Moderating role of core self-evaluations. *International Journal of Intercultural Relations, 43*, 177–188.
Xia, J. (2009). Analysis of impact of culture shock on individual psychology. *International Journal of Psychological Studies, 1*(2).
Yan, A., Zhu, G., & Hall, T. (2002). International assignments for career building: A model of agency relationships & psychological contracts. *The Academy of Management Review, 27*(3), 373–391.
Yuzhe, M., Soonkyoo, C., & Jaeyong, S. (2011). Transferring subsidiary knowledge in the global learning context. *Journal of Knowledge Management, 15*(3), 478–496.
Zeira, Y., & Banai, M. (1987). Selecting managers for foreign assignments. *Management Decision, 25*(4), 38–40.
Zikic, Z., et al. (2006). Repatriate career exploration: A path to career growth & success. *Career Development International, 11*(7), 633–649.

Chapter 3
Sectored Profile and Precursors Effecting Repatriation Adjustment

3.1 Introduction

International arena of business and a more global setting, increased international competition, improved communication technology and converging consumer habits and tastes have ushered in a globalized world of cross cultural interaction and management, wherein new markets for MNC'S have emerged. The consequent effects of this phenomenon of globalization of Indian Industry and its associated paradigms, the overseas operations of MNC's have gained significance in generating profits with over 50% of corporate revenues being generated away from the home country (Solomon, 2001). According to Stroh and Gregerson (1995), "Globalization refers to a process which deals in developing relatively standardized products with global appeal, as well as rationalizing the operations throughout the world and diversity, termed as "universal inclusion" wherein foreign assignments become a core component of the organizational and career development process". Subsequently, organizations are facing new challenges associated with cross cultural interactions and in particular the growing need of joint ventures, foreign subsidiary establishment, overseas representative offices and international strategic alliances has been accompanied by a combination of adaptive management practices.

The arena of Human resource management has become dynamic, leading to the critical component of strategy for the management of internationally mobile staff, demanding a revolutionized facet of human resource functions that caters to the specific demands herein. This has resulted in the emergence of IHRM (International Human Resource Management). IHRM is defined as the interplay between the three important dimensions of HRM—types of employees, HR activities and the country of operation and consists of three HR activities-procurement, allocation and utilization. The three major country categories involved in IHRM activities are– The host country where a subsidiary may be located and the employees therein (Home country nationals or HCN), the home country where the firm is headquartered and

M. B. Nag and F. Ahmad Malik, *Repatriation Management and Competency Transfer in a Culturally Dynamic World*,
https://doi.org/10.1007/978-981-19-7350-5_3

the employees therein (Parent country nationals or PCN), and third countries as a source of labor and finance (Third country nationals or TCN).

The growing aspect of internationalization of organizations in recent years has given impetus to the interest in the aspect of cross-cultural preparation and support of individuals undertaking International Assignments, and subsequently it has ushered in the relevance of Reverse Culture Shock and adjustment issues that face Repatriates. Poor repatriation is costly (Black et al., 2007, 1992), reduces the effective utilization of human resources (Adler, 1981), often leads to loss of talented personnel (Black & Gregersen, 1991a, 1991b), and probably represents the number one reason for employee hesitancy to accept overseas assignments (Harvey, 1989). The result is workforce attrition due to non-management of Repatriates returning to their home country after several years on a global assignment. In international business literature during the past 25 years, only more recently has repatriation begun to receive the same serious scrutiny as expatriation (Black, 1991; Black & Gregersen, 1991a, 1991b; Gomez-Mejia & Balkin, 1987).

The strategic targets addressed by an International Assignment are, to improve business performance, to foster the parent corporate culture in the subsidiary culture, to break the barriers between parent company and the subsidiaries and development of talent and future leaders of the company by opening up new international markets in a cost effective manner. Expatriate talent has manifold benefits as they play a crucial role for their organization (Tivey, 2008). Therefore there is a need of developing global employees with in-depth understanding of a global perspective. Corporate culture through assimilation and socialization, would therefore communicate goals and objectives in the manner in which they were intended since Expatriates and consequently the Repatriates, can enhance the competitiveness of firms.

3.1.1 Process of Expatriation and Repatriation: Perspective in a Globalized World

In research literature, the precincts of adjustment have been deliberated upon and domestic relocation and its manifestations have been posited by adjustment theorists (Ashford & Taylor, 1990; Brett, 1983), wherein the realm of Expatriation Adjustment (Black et al., 1991; Torbiorn, 1982) is akin to moving into a new environment and therefore creates a significant level of uncertainty whereby people in general have a need to reduce that uncertainty. Individual control theorists (Bell & Staw, 1989) argue individuals when placed in new and unfamiliar environments have a need to reestablish a certain level of control and this re-establishment of control primarily takes two forms-predictive control and behavioral control. The aspect of predictive control explains the ability to make sense of, or predict, one's environment in terms of the ability to predict how one is expected to behave and the ability to understand and predict rewards and punishments associated with specific behaviors (Bell &

Staw, 1989). On the other hand behavioral control is the ability to control one's own behavior that have an important impact on the current environment.

The holistic integration of relocation adjustment, expatriation and Repatriation Adjustment, and individual control theoretical foundations leads to the conclusive outcomes of adjustment which explains the fact wherein individuals establish behavioral routines based on their perceptions of expectations, reward and punishment contingencies, and their preferences for particular outcomes. Thereafter if they are confronted with new and unfamiliar situations, the established routines are broken and the individuals' sense of control gets reduced. Subsequently an individual attempts to reestablish a sense of control by reducing the uncertainty in the new situation through predictive and/or behavioral control, as per the need of the situation or by default of perception (Bell & Staw, 1989).

In the contemporary globalized world, the organizations are facing new challenges associated with cross cultural interactions. The growing need of joint ventures, foreign subsidiary establishment, overseas representative offices and international strategic alliances has been accompanied by a combination of adaptive management practices and hence the changing interdependence between IHRM and other corporate functions (Mendenhall, 2007). Though, there has been considerable research undertaken on the aspect of selection and training of expatriates, few have studied the last part of International Assignments—the Repatriation Process (Black, 1991; Solomon, 1995). A major reason for the repatriation process not to be seen as an important issue is also because of the common opinion that—the expatriate is only coming home and that this should not cause any problems. However, studies have shown that repatriating is often more challenging than expatriating (Black & Gregersen, 1991a, 1991b). Research indicates that there are several problems, covering the practical and psychological realms of the repatriation process (Sussman, 2000).

Lazarova and Caliguri (2001) contemplated that Repatriates become more marketable due to the enhanced skills they have developed during an International Assignment such as language proficiency, differing management techniques and perhaps more importantly a differing global view. Thus retaining the newly acquired skills and knowledge from the Repatriate after the assignment should form a part of the success criteria equally important to the individual and organization. Adjustment of Repatriates is crucial to the corporate and employees on the financial as well as on the reputation front.

Though the strategic and financial importance of repatriates is unmistakable, lot of evidence indicates that MNCs often fail to capitalize on these human investments, because many of their assignees leave the company after the assignment is completed, due to; Being placed in non challenging jobs, Lack of promotion opportunities, Loss of status and autonomy, lack of career planning and counseling, Lack of support on behalf of managers and colleagues, Sluggish career advancement and Repatriates' perception of how well the MNCs managed their repatriation process. Given this strategic human capital issue, ways to predict Repatriate retention and lower their turnover is an important challenge facing organizations (Black et al., 1992, 1995; Stroh and Caligiuri, 1997). This is supported by numbers as according to 2005, Global

Relocation Trends Survey, issued by Global Relocation Services and the National Foreign Trade Council, 23% of repatriating employees leave their company in the first year. After three years the total is at least 40%, though some people place it closer to 50%. These figures represent not only widespread disruption in the lives of employees and their families but a staggering loss of corporate investment.

Given the scenario, Human resource departments, in particular, have great responsibilities to ensure a smooth re-adjustment for employees returning to home country. However, trends reveal that only a minority of companies invest substantial resources in this last task of creating an Effective Repatriation process, even though researchers have confirmed repatriation to be more challenging than Expatriation (Mooradian, 2004). The cost of failed International Assignments is high, both financially for the organization and from an employee's perspective (Webb, 1996). The cost of repatriation failure is threefold (Goss and Hynes), state that the first area is "underutilization of key employees" and the second is the "loss of key employees". Another area is the "inability to recruit employees into overseas positions", and therefore decreased cross-cultural knowledge and understanding. As per Black and Gregersen (1991a, 1991b), many expatriates are dissatisfied with the repatriation process. This paved way for researchers to examine the factors affecting the repatriation process from an HR perspective (Mendenhall & Odou, 1985), which included; Family difficulties posed by assignments, Selecting candidates that are either unsuitable or unwilling to accept assignments, Not taking advantage of the skills and knowledge acquired by assignees, Losing employees after repatriation, Remaining competitive while trying to control costs and provide consistent policy, coverage, Security of assignees and non-compliance to various laws and regulations by assignees, to name a few.

Herein, success in terms of achieving the organizational aims as well as the individuals' terms for success are hardly analyzed and are therefore under researched. Therefore, the factors that influence organizations to utilize repatriation programs, and the function of international transfer and organizations' beliefs about the value of international experience will determine the scope of repatriation program. There has been a lot of debate about the corporate imperative to support its expatriates and their families. However, as research proves, less has been analyzed regarding, what befalls these expatriates when they return home to pick up where they left off and therefore necessitates more holistic research.

3.2 The Concept of Repatriation Management and Its Associated Research Paradigms

The above cited dynamic parameters of IHRM have brought forward the nuances of handling repatriation effectively and therefore the concept of "Repatriation Management" in the domain of cross cultural management has evolved. The initiatives that capitalize on the re entry of a firm's workforce, comprise of those strategic HR tools that handle Reverse Culture Shock, optimize talent, ensure smooth transition from being an expatriate to a Repatriate, Diaspora option, debriefing sessions for reintegration in home office, family adjustment and re-acculturation, HR initiatives for career development and profiling from an overseas assignment, as a strategic approach to business that contributes to the organizational goals of productivity and profit (Black & Gregersen, 1991a, 1991b). Several areas of research are now focusing on the analysis of Expectation theory of Repatriates, Re-entry systems theory, Equity theory and The "W-curve theory". These studies strive to bring in tandem the dimensions of acculturation of Repatriates and the necessitated benchmarks that have to be kept aligned both by the individuals and the organizations.

The issue of differing expectations between the assignee and the organization has less widely been discussed and research literature pertaining to management of Repatriation, has started pouring in recently (Black & Gregersen, 1991a, 1991b; Hyder & Lovblad, 2007). In the said context, Paik et al. (2002) explained that discrepancies often exist between the motivations and expectations of HR managers and assignees. Various dimensions studied thereafter, included, timing of the role and negotiations as adjustment facets (Suutari & Valimaa, 2002). It has been contemplated upon, that the overall responsibility for repatriation activities should arguably rest with the home sponsor who maintains an interest in the assignee from the outset of the assignment (Lazarova & Caligiuri, 2001). The framework of study considers uncertainty reduction, as a key concept to further research into International Assignments. In addition, organizational commitment was considered important during the repatriation process in order to maintain a positive relationship to aid retention. The introduction of social cognitive career theory widened the research agenda by contrasting the theory of Uncertainty Reduction Theory, wherein Tharenou, (2002) suggested that individuals take up expatriate experiences in order to create uncertainty and thus are receptive to international work experiences. This was an anti-thesis to the previous research that focused upon reducing uncertainty, as a prelude to repatriation. Whilst academic debate had been proposing plausible models to aid the enhancement of repatriation, there was little evidence of research to test such models (Taveggia and Santos, 2001). In this aspect, the longitudinal work of Suutari and Brewster (2003) was one exception which tracks cohorts of expatriates from their assignment to the return home. The results from this study indicated that many Repatriates leave the organization once repatriated thus repatriation offers a critical challenge to organizations (Bonache et al., 2009).

The various aspects of Repatriate adjustment and management are essentially in the domain of psychological adjustment, socio cultural adjustment and the self construal of individualism or collectivism as exhibited by the Repatriate Therefore research since the early nineties delved deeper into the concerns of Repatriation and its management, bringing out several criterions resting on the periphery of Human Resource Management practices till then. In recent times, the concept of Repatriation Management has achieved enormous importance, given its applicability in the International arena at large.

3.2.1 Successful International Assignment: Precursors and Outcome

As a result of increased International Assignments, the variables of success and what it implies for the organization and the individual becomes more significant in each of the phases of an assignment including the final repatriation stage. The parameters of a successful International Assignment are variable and so are the reasons for the high failure rates of assignments. Welch et al. (2007) proposed that the main reasons assignments prematurely end comprise; the partners' dissatisfaction; inability of the assignee to adapt; family adjustment; differences with managerial styles; culture and language difficulties and partners' dual career issues. Failure can be defined as, the inability of an expatriate to perform effectively in a foreign country and, hence the need for the employee to be fired or recalled home (Tung, 1998). This failure is related to the costs associated with premature return as well as retention upon repatriation (Allen & Alvarez, 1998; Howe-Walsh, 2010). Other factors for success such as knowledge transfer (Paik et al., 2002) and performance on assignment are less researched upon. Rieche (2012), explains that workers with highly specific skills are less likely to quit their jobs. Suutari and Brewster (2005) suggested that an assignment is one of the most effective methods to provide a learning opportunity for the organization. Repatriation is a key turning point in an employees' career as they are vulnerable to leaving the organization at this stage (Vidal et al., 2007). It was analyzed that the employment status including reduced work status had caused significant difficulties upon return, resulting in potential failure for the individual and the organization in terms of their employment (Black et al., 1991).

Furuya et al. (2007) elaborated that success includes readjustment with the Repatriates' work circumstances as well as commitment to their work and employer. Other factors related to the assignment which comprised whether the assignment fulfilled any other criteria such as knowledge transfer (Lazarova & Caligiuri, 2001; Suutari & Brewster, 2003) seldom remain the focus but require equal consideration. From the organizations' perspective, should a Repatriate leave the organization within twelve months, the assignment can be deemed a failure, as a loss of knowledge and skills is the consequence of such attrition (Suutari & Brewster, 2003). On the contrary, from the individuals' perspective, leaving to work in another organization could be beneficial to their own career (Black et al., 1991). Lazarova and Cerdin (2007) posit that not all turnovers are dysfunctional and that some assignments are not of strategic importance to the organization and are thereby used for limited term positions.

This implies that the success of an assignment is difficult to conclusively define and therefore the success for an assignment requires consideration of the initial objectives which should incorporate planning for repatriation. From a broader perspective, success may be defined as an assignment that aids in the development of an individual, lasts the prescribed time and leads to the optimization of the individual and the organization. Influences upon success of an assignment such as HR policies and practices as well as whom and how these are delivered, require further research in order to contribute to the knowledge of repatriation.

3.2.2 *Framework of Repatriation Adjustment*

The conceptualization of the aspects of Repatriation Adjustment have primarily focused upon the adjustment to work (Brett, 1984) while the expatriation literature has primarily focused on adjustment to the culture of the host country (Mendenhall and Odou, 1985). However the trio of Black et al. (1991) conceptualized expatriation adjustment as a multifaceted phenomenon which comprises of three related yet separate facets: adjustment to work, adjustment to interacting with host nationals, and adjustment to the general environment and culture (Black & Gregersen, 1991a, 1991b). The mentioned studies found that while some variables affect all three facets of expatriation adjustment, other variables only have a significant impact on one facet. Black and Gregersen (1991a, 1991b) took this research ahead and found evidence that Repatriation Adjustment also has three related yet distinct facets: adjustment to work, adjustment to interacting with home nationals, and adjustment to the general environment and culture.

On the contrary successful repatriation means that the Repatriate expects and acquires a feasible career and personal development after his International Assignment. As cited above, research has proven that this does not happen to all Repatriates after an International Assignment. According to Geodesy survey, 33% of Repatriates are promoted, 58% of Repatriates stay at the same level and 9% are demoted after returning from International Assignment. Further, when the Repatriates return to the home country, there is a general assumption that the Repatriate will be able to settle quickly into the community. However many Repatriates face Re-entry shock or Reverse Culture Shock as individual expectations before repatriation, during expatriation and after repatriation do not match the organizations' expectation in terms of position, level of income etc. This lack of understanding to integrate repatriates into the organization in terms of the knowledge and experience they gained during International Assignment is a factor that often leads many Repatriates to quit the organization (Black et al., 1991).

This has been reiterated by Black et al. in 1992, wherein it was argued that individuals make anticipatory adjustments before they actually encounter the new situation. Furthermore the importance of accurate expectations to facilitate adjustment was analyzed. From the international adjustment literature three categories emerged regarding pre-departure variables important to adjustment, namely, previous experiences, pre-departure training and candidate selection. Two variables, related to post arrival, include individual skills and non-work factors. Overall adjustment is referred to in terms of adjustment to the home culture (Black et al., 1992). Black et al. (1992) developed a theoretical framework to address the area of Repatriation Adjustment. They argued that the area of repatriation offers additional challenges and that developing a systematic approach can assist with predictors of Repatriation Adjustment. Black and Gregersen (1991a, 1991b) found that evidence to suggest that Repatriation Adjustment also has three related yet distinct facets: adjustment to work, adjustment to interacting with home nationals, and adjustment to the general environment and culture. Thus they suggested that Repatriation Adjustment is multifaceted which is important for two reasons. Firstly an antecedent such as the support of co-workers could be related to one or all facets and secondly, if Repatriation Adjustment is multi-faceted then is it possible that certain facets will be more or less strongly related to outcomes such as turnover and job performance (Black et al., 1992). The Repatriation Model (Black et al., 1992) is divided into two parts; pre return adjustment conducted prior to repatriation and post return adjustment. The model groups variables into: individual; job; organizational; and non work. Within each set of variables several relating factors are highlighted. Pre return adjustments include how realistic the Repatriates' expectations are from an individuals' perspective. Whereas organizational variables cite communication with the home location, which is clearly central to the amount of information available regarding future return positions etc. Additionally the Repatriate would seek other sources of information related to non work variables again to set expectations for what it will be like to return. Post return adjustment brings together the same variables but offers discussion as to how the Repatriate is likely to react to changes to their home work environment. This raises the issue of maintaining the Repatriates' momentum during repatriation process to capture the enthusiasm facilitated by an assignment and lead to a positive experience of the repatriation process (Fig. 3.1).

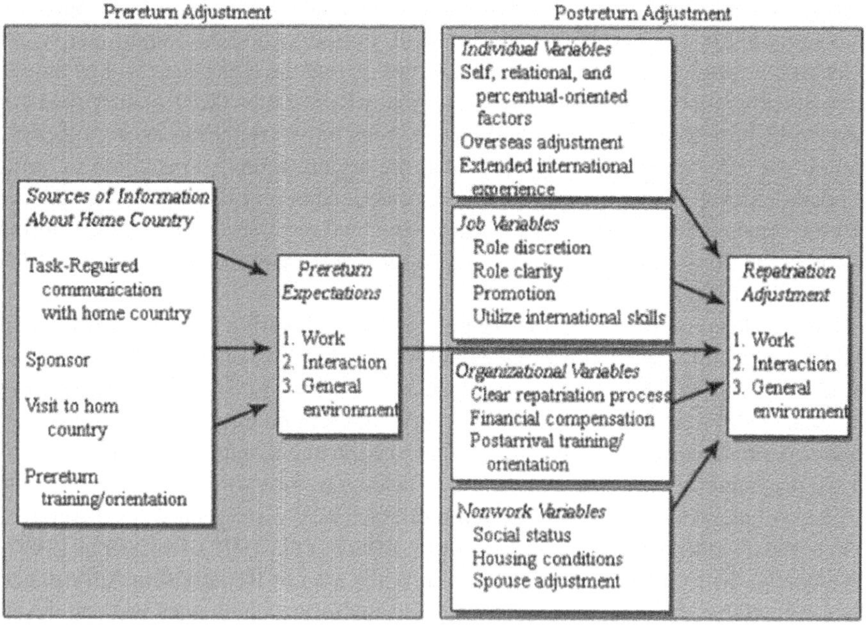

Fig. 3.1 Basic framework of repatriation adjustment. *Source* Black et al. (1992)

3.3 Paradigms of Repatriation Adjustment Proposed in the Current Study: Background and Research Implications for International Human Resource Management Practices

3.3.1 Factor 1: Competency Transfer

The aspect of transfer of knowledge upon return, is one of the most important strategic objectives related to successful Repatriation Management. Expatriating individuals gain not only substantial knowledge related to particular, local management practices and local market conditions, but also develop a number of overall management skills and interpersonal relations competencies (Lazarova & Tarique, 2005). Unfortunately, Re-entry management is rarely high on the HR agenda of multinational corporations (Bonache et al., 2001) and the returning individuals are confronted with bottlenecks when attempting to transfer their newly gained foreign knowledge to the home-organization. Adler (1981) labels this negative reaction a xenophobic response. An unfavorable attitude among home-country employees is found to hinder the transfer of knowledge from the host- to home-country organization and negatively influence the re-adaptation of the returning individuals. This may be a resultant effect of twin strategies of a Repatriate masking the source of their knowledge and, in this way, prevent a prejudiced response. Secondly, the home-country managers not offering assistance (training) in recognizing the benefits of the international experiences their Repatriated colleagues have gained. Studies have reported that expatriates receive significant personal and self-development benefits from foreign assignments. The major gainful areas are with respect to, gaining a broader, global perspective on the firm's operations and enhanced communication skills as a result of exposure to different cultures. The more complex the environment, the more it enhances their planning and motivation techniques, thus, confidence increases. These types of development can have positive career and organizational implications.

In the present study, Repatriate Competency Transfer is taken into consideration, while analyzing the antecedents of Repatriate Adjustment. The hypothesis corresponding with this is pertaining to the aspect, whether Effective Repatriation is a significant predictor of Competency Transfer of a Repatriate. It comprises issues pertaining to the transfer of competencies of Repatriated employees per se. Research suggests that if International Assignments were designed keeping in mind the different aspects of skill and acumen transfer principles, as elucidated in the variables under the construct of Competency Transfer, Repatriate talent would get optimized. It signifies that Repatriates who are able to make use of their overseas acquired acumen, resulting from their international experience tend to feel more satisfied with their career growth. This would suggest that those organizations that identify a specific career track flowing from a foreign assignment will have a more satisfied Repatriate work force (Naumann, 1993). It has been seen that expatriates demonstrate a high commitment to the international element of their careers and to its value as a means of securing organizational success into the future (Suutari, 2003). Bonache (2005) in a study comparing satisfaction of expatriates of both domestic employees and Repatriates found that expatriates are in general more satisfied with a number of job characteristics. Precisely, the extent to which their jobs have task variety and autonomy, provide opportunities for learning, allow them to apply their knowledge and give them enough responsibility, are the favorable aspects which are looked forward to, on return. These findings lend support to authors such as Suutari (2003) who claim that the best part of International Assignments is related to the new work experiences and learning they offer.

According to Lazarova and Tarique (2005), for an effective transfer of knowledge upon repatriation to take place there must be a fit between both the organization's interest and openness, and the Repatriate's readiness to share the information. Proactive organizations could also undertake a number of actions supporting knowledge transfer (Lazarova & Caligiuri, 2004). Such procedures should be organized in the form of anticipatory policies rather than being implemented as ad hoc crisis management interventions (Lazarova & Caligiuri, 2001). Research indicates the importance of appropriate HR practices and self-adjustment efforts of returning individuals contributing to 'within-company learning processes' (Furuya et al., 2007). It is desired that HR managers devise practices to better manage this critical resource (the expatriates) in order to maximize the wealth of knowledge and experience gained from International Assignments. Performance development and management practices would have to be designed in such a way so as to capture the best of development opportunities (Fenwick, 2004; Webb, 1996).

3.3.2 *Factor 2: Effective Repatriation*

This factor group comprises the Effective Repatriation training and socialization methods employed by parent organizations. It is imperative that not just technical considerations but also an employees' willingness and motivation to relocate, albeit temporarily, are not ignored. Cross cultural training initiatives, realistic job previews and subsequent socialization practices are of pivotal importance in this regard. Given that IT projects cover a large range of technical knowhow (and roles ranging from developers, analysts, system architectures and managers) paying heed to their adjustment back home with respect to in-country and organizational adjustment is important. The strength of this factor is that it helps to gather information from various sources in order to understand what needs to be done to improve and optimize Repatriate talent and its subsequent adjustment back home.

This factor group addresses the issues pertaining to cross cultural adjustment both in the workplace and in daily life. Feelings of stress in a new work environment, ability to decode the communication (verbal and non-verbal) and social etiquette and acclimatizing oneself with the way work is done in terms of interpersonal and team protocols comprise this set of factors which impact satisfaction, a Repatriate feels on his return back home. This issue is enhanced by the "Out-of sight, Out-of-mind" attitude that most firms take with respect to expatriate employees (Adler, 1981; Harvey, 1989). This change manifests itself in the fact that changes in the individual, which adjustment to the foreign country while overseas might foster; and changes back in the home country and home office would be more important factors for Repatriation Adjustment. Based on the difference of degree and difference in kind discussed, it would seem that while the domestic relocation and Repatriation Adjustment process would not be totally unrelated to each other in terms of the theoretical process, however they are sufficiently different wherein some variables that play a minor role in domestic relocation adjustment (Job, Organization and Environmental novelty) play a critical role in Repatriation Adjustment. In addition, demographic variables, such as "time away from the home country," that are irrelevant in domestic relocation adjustment, would be of concern in Repatriation Adjustment. This interplay of variables demonstrates sufficient differences to warrant separate theoretical and empirical investigation of Repatriation Adjustment.

Repatriate adjustment has been commonly defined as "the degree of comfort or absence of stress associated with being a Repatriate, (Shrinivas et al., 2005). Black et al. (1991) provide a framework for the three different levels of adjustment Repatriates experience. General adjustment is the degree of comfort with living conditions, climate, food, housing and cost of living. Interaction adjustment is the individuals' comfort at socializing and interacting with home nationals. The third dimension is adjustment to work, which includes understanding and adapting to specific work duties, responsibilities, standards and expectations. Repatriates reach a level of comfort when all levels of adjustment have been satisfied. Studies have revealed several factors associated with Repatriates" improved socio-cultural adjustment (White et al., 2011).

In this context, the role of the extent and nature of support from the employees' own organization cannot be ignored. Maintaining feasible and appropriate communication channels with the newly assigned expatriates and continuous monitoring of events may provide a basis for assessing their emerging support needs (Selmer et al., 2007). It is important that expatriates need to be more tolerant and optimistic in their attitudes towards cultural differences that occur in a cross-cultural environment (Tahir and Ismail, 2007). It has been observed that cross-cultural training affects the accuracy of expatriates' expectations prior to the assignment—and that having accurate expectations, in turn, positively affects cross-cultural adjustment (Caligiuri et al., 2001).

3.3.3 Factor 3: Anticipatory Adjustment

There is a lack of systematic investigation of Repatriation Adjustment and this may stem from an underlying assumption that the process of Repatriation Adjustment is not significantly different from adjustment after a domestic relocation transfer or from adjustment to an overseas assignment. This may stem from the frequently held assumption that both home-country and home-organization are well known entities and do not need further explanation. A 'reality check', however, often reveals that both the home environment and the returning individual have changed substantially during the period of intercultural sojourn (Black et al., 1992).

Although Repatriation Adjustment may be related, in some ways, to other incidents of adjustment, it is a significantly different phenomenon and is worthy of separate theoretical and empirical interest and investigation, as has been proven by research related to Repatriation. General perception related to Repatriation Adjustment is that it is the result of a job-related geographic transfer; however it is important to distinguish it from the two other major types of job-related geographic domestic relocations and overseas assignments. Unlike the individual who remains in the home country during the same period, the repatriating employee is unable to examine the interplay between his or her personal changes and the environmental changes back in the home country which may affect Repatriation Adjustment. These personal and environmental changes suddenly confront each other upon the individual's return to the home country. This sudden confrontation can bring forth some critical factors giving evidence, as to why most employees report that reentry into their home country and company was more difficult than the initial move overseas (Adler, 1981).

This implies that Repatriation Adjustment framework theoretically consists of both adjustments made prior to returning home, namely Anticipatory Adjustment and adjustments made after arriving back home, namely In-country Adjustment. Relocation, Expatriation, and Individual control literature suggests that there are four main categories of variables which impact such movements (Glisson and Durick, 1988) Individual Variables that are a function of individual attitudes, values, needs, or characteristics, Job Variables that are related to the tasks and characteristics of the individual's job and Organizational Variables that are related to the organization and its policies and practices. The category of non-work variables would comprise of variables that are traditionally outside the work and organizational domain such as the family and the general environment.

Individuals would not necessarily make actual behavioral changes before, and in anticipation of, returning to their home country, but it is likely that the anticipatory changes would consist of cognitive adjustments. Repatriation Adjustment conceptualizes three facets (Black & Gregersen, 1991a, 1991b), which suggests that anticipatory expectations of greater significance are concerning work, interacting with others, dealing with the general environment back home. Individuals engage in efforts of predictive control or trying to form and modify expectations about their new job, social interactions, pertaining to dimensions back home. Adjustment theorists (Ashford & Taylor, 1990) argued that accurate expectations will in general facilitate adjustment to the actual circumstances. This is because accurate expectations reduce the chance of "surprises" and their associated uncertainty. Bandura (1991) posited that specific expectations would be most strongly related to their corollary, specific outcomes, and that general expectations would be most strongly related to general outcomes. Thus, one would expect accurate work expectations to have their most positive relationship with in-country work Repatriation Adjustment; accurate interaction expectations to have their most positive relationship with in-country interaction Repatriation Adjustment; and accurate general expectations to have their most positive relationship with general environment Repatriation Adjustment.

Expectations Model provides a detailed explanation of the clash between individuals' expectations of reentry and the reality of the transition (Adler, 1981; Black, 1992). It elucidates the pre-reentry expectations and their influence on the process of re-adaptation to the home-country environment. Similar to the Reverse Culture Shock model, this theory was initially developed to address the expatriation/outbound experiences familiar to the cross-cultural field (Black and Gregersen, 1990; Caligiuri et al., 2001). However, while outbound difficulties and the need for preparation prior to expatriation were acknowledged relatively quickly by scholars and accepted by travelling sojourners, the discomforts associated with reentry still remained largely unexamined and unexpected (Black et al., 1992).

Investigations on the influence of expectations on the repatriation process have been conducted, but no major consensus has been arrived at (Stroh et al., 1998; Sussman, 2000). MacDonald and Arthur (2003) claimed that met and exceeded expectations lead to a smooth readjustment. Meanwhile, in a job-related context, Stroh et al. (1998) advocated that, while met expectations support Repatriates' commitment and intention to stay in their home-country organization, exceeded expectations might lead to an even higher dedication to the workplace. Black's (1992) concluded that accurate expectations result in the highest level of readjustment and job performance. Consequently, Black et al. (1992) argued that the readjustment process should be made not only after reentry, but, most importantly, it should include cognitive anticipatory adjustment prior to return. This would help in building up a realistic level of expectations. These conclusions were supported by Maybarduk (2008).

Factors that influence uncertainty and loss of control are therefore expected to be most relevant in the general adjustment process. Factors that reduce uncertainty would facilitate adjustment, while those factors that increase uncertainty would inhibit adjustment. This sums up the psychological and situational aspects of adjustment. The fundamental process of adjustment encompasses an important dimension which deliberates upon the fact that that unless the event of entering the new environment is totally without warning, individuals can and do make anticipatory adjustments (Black and Mendenhall 1990; Black et al., 1991). Therefore, adjustment as a phenomenon signifies the fact that individuals can either imagine or simulate the new environment before actually entering it and that they aspire to reduce both the uncertainty and perceived loss of control in anticipation of the move, thereby forming expectations.

3.3.4 Factor 4: Reverse Culture Shock

International Assignments are not permanent relocations and at one point the employees return to their parent organization back home. In this context, it is important that the expatriates are given ample notice and opportunity for a phased transition to deal with the Reverse Culture Shock that they may face after being on International Assignment for a prolonged period. Issues pertaining to organizational attitude towards extensions/renewal of work Visas, rotation of employees on international projects and abrupt departures and transitions and opportunities of growth post repatriation and comprise this factor. Repatriation concerns are largely based on how expatriates perceive their colleagues (past expatriates) experiences post repatriation. Adjusting to repatriation is often more difficult than adjusting to life overseas, as the Repatriate may have great aspirations on the career front as a result of acquiring global exposure, skills and experience.

Another issue in this context is how the assignment came to an end. Also strategic policy changes in the client organizations may have an impact on tenure and nature of the assignment. Added to this, is the fact that on repatriation they have lost most of the generous financial benefits which is associated with expatriate assignment, and because there are often no plans for repositioning them within their organization, they are made to feel as if their assignment was of little or no value to management. They might be expected to resume their earlier roles because management does not know how to use their expertise. Expatriate assignments, which many believe will increase career mobility, can actually slow down promotions. Repatriation experience is dependent on two main factors: how the reality of the repatriation process meets the expectations of the Repatriate; and the impact of demographic factors and cultural identity changes (Hyder & Lovblad, 2007; Webb, 1996). Human resource professionals should conduct exit interviews with expatriates and their families on their return home. There is an abundance of practical knowledge and applications to be learned during an expatriate assignment that can be useful to future expatriate workers. Some issues on re-entry include xenophobic attitudes of colleagues and managers at home, underutilization of global managers, unplanned transition and career paths. Organizations should reflect on are how they might help expatriates manage the re-entry transition and make a returnees' transition as easy as possible for the Repatriate and as beneficial as possible for the company (Adler and Gunderson, 2008).

Returning employees get wary of the aspect regarding the way the international assignments will impact their subsequent career development and whether their career development expectations will be met (Yan et al., 2002). This can be countered by the organizations by preparing the returning expatriate with an orientation and adaptation program, enabling their preparation for the changes that can impact their expectations and subsequent work adjustment. Black et al. (1992) recommend a regular exposure to the home country and domestic office to facilitate accurate work expectations. Permitting the employee to negotiate various new job assignments allows the Repatriate greater control over returning to the home country organization and helps facilitate adjustment. The need for behavioral control and self-efficacy can be met by creating opportunities that incorporate Repatriates' international competencies into career trajectories (Feldman & Tompson, 1993).

3.4 Adjustment and Reverse Culture Shock: Process and Outcome

The concept of "Cross-Cultural Adjustment" began with earlier work on culture shock. The term Culture shock was introduced by an anthropologist, Dr. Kalervo Oberg in 1960, to describe the trauma suffered by people when they are suddenly introduced to a culture that is very different from their own. In 1988, Befus, described culture shock as "An adjustment reaction syndrome caused by cumulative, multiple and interactive stress in the intellectual, behavioral, emotional and physiological levels of a person recently relocated to an unfamiliar culture and is characterized by a variety of symptoms of psychological shock. Oberg (1960) described four stages of adjustment. Oberg introduced the term 'Culture Shock' in order to describe the experiences related to living in an environment that differs culturally from the settings one grew up in.

He described the challenge of returning home being related to issues of unexpectedness of the difficulties encountered and a lack of preparation for reentry, besides the grief for the lost expat life. Individuals can display considerable emotional distress as much as six months after their reentry. Research indicates that the extent of this distress can reach clinical levels (Chamove & Soeterik, 2006). The intensity of grief experienced by returnees can even be comparable to that experienced by bereaved individuals. Issues of psychological distress have been widely documented among different returnee groups; amongst students (Gaw, 2000; Szkudlarek, 2010; Thompson & Christofi, 2006) amongst International assignees (Black et al., 1992b; Furuya et al., 2006) and amongst refugees (Sundquist & Johansson, 1996). Nevertheless, there is a contradiction in the research describing the challenges of reentry are not unanimous. Sussman (2001) reports that the migrants returning to Hong Kong did not exhibit the negative affect associated with return migration. These results indicate a need for more diversified research on affective aspects of reentry transition.

Basis this explanation and its associated nuances, the phenomenon of Reverse Culture Shock has been categorized into processes of adjustment of a Repatriate, beginning with his acculturation and adjustment process upon his sojourn, as an Expatriate and culminating upon his return to the home country, over a period of time, during the course of an International assignment. Two main theories of Repatriation Adjustment have been propounded, basis the stages manifested in this cyclical sojourn.

A. **U-Curve Theory of Adjustment**

Early research on adjustment focused on the U-Curve Theory (UCT) of adjustment. The U-curve model describes four distinctive phases of an intercultural sojourn. This theory represents the four stages of Repatriate Adjustment.

I. *Honeymoon Phase*

It is the period where expatriates will have initial enthusiasm to interact with host nationals. There is euphoria or positive excitement about a new environment It is a period lasting less than two months (Adler, 1986; Black & Gregersen, 1991a, 1991b; Gullahorn, 1962; Torbiorn, 1982).

II. *Culture Shock Phase*

The Repatriates have to cope up with the new culture on a day-day basis. In this stage, the sojourner lacks the understanding of an alien culture, inhibits awareness of what is appropriate or inappropriate behavior in the new cultural environment, resulting in confusion, frustration, tension and depression (Adler, 1986; Black & Gregersen, 1991a, 1991b; Brislin, 1981). Hence, this is the Culture shock caused by mostly negative or surprising, experiences in a new environment.

III. *Acculturation Phase*

It is characterized by increased ability on part of the Expatriate to 'fit in' in the new culture. Acculturation or the learning process of adaptation to a new environment, takes place. Individual traits, cognitive and behavioral aspects of his personality may play a role herein.

IV. *Adjustment Phase*

Here, adjustment is almost complete and anxiety is over and a stable state is achieved if the acculturation process is accomplished successfully In short, an individual deputed on an International Assignment may first feel excited, then face problems and finally adjusts over time to the new socio-cultural environment.

Subsequently, researchers have found that not all sojourners experience the same level of anxiety, or experience anxiety for the same length of time (Church, 1982). This resulted in the study of cross-cultural adjustment as an individual difference criterion, which could potentially be predicted, rather than as a fixed period of anxiety that all sojourners will necessarily experience when they enter a new culture (Black, 1990).

B. **W-Curve Theory of Adjustment**

The W-curve is an extension of the above-mentioned model and describes the four previously mentioned elements: Euphoria, Culture Shock, Acculturation and the Stable state, experienced by a Repatriate after his/her return to the home-country (Fig. 3.2).

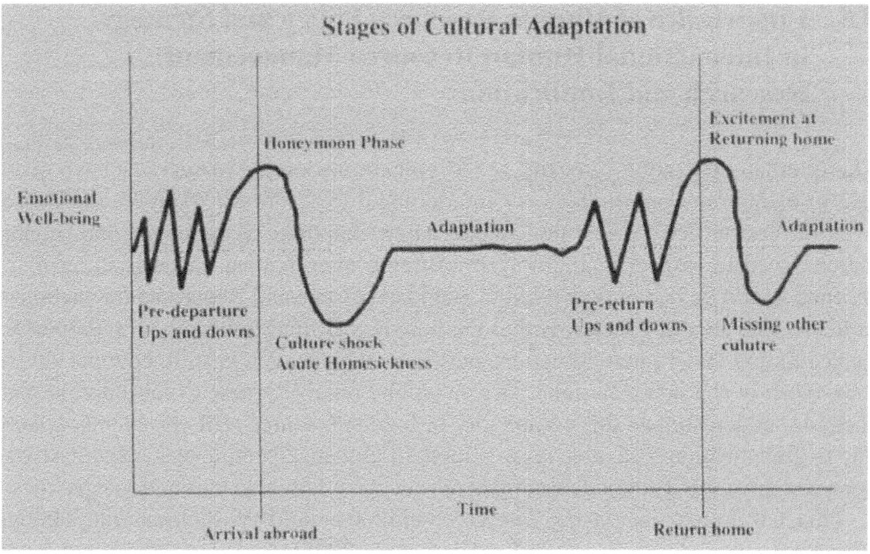

Fig. 3.2 W-curve theory of adjustment. *Source* Baloch (2013)

The work of Gullahorn and Gullahorn (1963) and their Reverse Culture Shock model, also called the W-curve theory, is a hallmark study in the realm of Reverse Culture Shock. Their model is a theoretical extension of the U-curve hypothesis. Repatriation Adjustment issues were identified in the literature in the 1950s and 1960s (Gullahorn & Gullahorn, 1963). Adler (1981) discovered that employees found reentry into their home country and home company more difficult than the initial move to the foreign culture. Once the individual returns to the home country, the process of in-country Repatriation Adjustment begins.

Gullahorn and Gullahorn (1963) argued that the cross-cultural adjustment process which led to a U-Curve can be applied to Repatriation process as well. Harvey argues that the longer one spends away from home; more changes can happen in home country. This leads to greater uncertainty upon return. The literature review of culture shock indicates that an individual may not function effectively during the period of cross-cultural adjustment. Black et al. (1992) stated that the two most affected work areas of Repatriation Adjustment are performance and turnover. Based on their research, they found that failing to pay attention to Repatriation Adjustment may lead to low intention to stay with the firm (Black et al., 1999).

These four stages are accompanied by strong affective responses, which in turn influence Repatriates' readjustment to their home-country environment. Black et al. (1992) reported that nearly 70% of the returning individuals cited significant discomforts related to their repatriation, which also indicated that psychologically challenging parameters of the repatriation phase were more severe than those of the expatriation phase (Adler, 1981; Chamove & Soeterik, 2006; Linehan & Scullion, 2002).

3.5 Importance of Human Resource Policy and Strategy in International Human Resource Management: Research and Implications

The Internationalization of business and global outsourcing of services have given rise to mammoth opportunities for international exposure for today's' workforce. Multiple factors have discerning impact upon expatriate assignments and organizations need to respect that simply prioritizing international business in terms of revenue would be incomplete without sound expatriate and Repatriate management policies. The area of concern is that the human and financial costs of underperformance/failure in the international business area which can relatively be more severe than those in the domestic area. Organizations relatively new to the international scenario, underestimate this complexity of human resource (HR) problems, arising due to globalization. Effective management of globalized employees is increasingly been recognized as a major determinant of success or failure in international business.

This has led to increasing interest within the field of International Human Resources (Brewster et al., 2007; Dowling et al., 2008). However there is not much clarity with regard to the implication for the delivery of HR policy and practice leading to the effective management of international staff (Lazarova & Cerdin, 2007). Friedman (2007) gave a theoretical review of the implications of globalization for HRM roles. However empirical research is scant. Research indicates that an HR department can add value and provide a co-ordinated approach to the activities that will assist the assignee and their family to pursue an International Assignment (Bossard & Peterson, 2005; Lazarova & Cerdin, 2007). This involves facilitation of how the assignment is managed from initial selection to how the eventual repatriation becomes central to the discussion of effective management of the International Assignment. If the HR manager has experience of an International assignment, there is an increased chance that the process of handling Repatriation issues is more effective (Suutari & Brewster, 2003). However, research indicates that approximately only 11% of HR managers have worked abroad (Black and Gregersen, 2007) which explains, why line managers are often left to negotiate the assignment terms on an informal basis. Research refers to the cultural adjustment training (Black et al., 1991; Tharenou, 2002) and there is little evidence of research into the actual HR policies and practices that support the effective management of the expatriation process including repatriation activities (Bossard & Peterson, 2005; Lazarova & Cerdin, 2007). According to Black and Gregersens' (2007) research, the HR function and line manager are expected to lead the repatriation process with the Repatriate. Part of this process would entail debriefing the Repatriate via an outsourced function. An outsourced function, based in the same country, involves administering parts of the HR function such as arranging removals, briefing assignees before their assignment etc. to a third party (Sparrow & Braun, 2007). This deliberate management of the process is where the HR function can add value to the overall strategy of the business by supporting the repatriation activities.

In order to progress our understanding of the effective management of repatriation processes it is useful to explore the role of HR practices and policies and their ultimate delivery. However, there is scant regard within the literature relating specifically to HR policy and practice for repatriation. A study related to the HR function specifically related to recruitment and selection conducted by Sparrow and Braun (2007) highlights the gaps between policy and practice. The study argues that there are variances between macro-level policy and what happens when the processes are converted into the perceptions of the end users e.g. the employee or HR manager. The lack of research to test the effectiveness of practices and policies is even more evident within the expatriate literature that appears to preclude research into HR policy and practice for International Assignments (Lazarova & Cerdin, 2007).

Indeed how the HR services are provided to the organization will impact upon the roles HR plays within the process which appears to be precluded in the literature. Therefore, research into the role the HR department plays will increase the understanding of how it can add value to the process and further our understanding of its contribution towards successful repatriation, as well as fill the existing gap within the current academic literature. Thus, it is undeniable that a research into the area of HR practices at an organizational level by developing a greater understanding of repatriation processes and supporting HR policy is called for.

3.6 Corporate Repatriation Management: Indian IT-Sector Perspective

3.6.1 Corporate Repatriates

The most thoroughly researched group within the reentry field, remain the Corporate Repatriates. This is in partly due to the fact that work-related difficulties are among the most often reported concerns upon return (Adler, 1991; Baruch & Altman, 2002; Black et al., 1992; Furuya et al. 2007; Kraimer & Wayne, 2004; Linehan & Scullion, 2002; Paik et al., 2002; Stroh et al., 2000). Issues related to workplace readjustment, commitment, transfer of knowledge, productivity and post-reentry turnover rates among returning individuals have been widely discussed in the international management literature. Lazarova and Caligiuri (2004) argued for the individual treatment of different corporate assignee groups (technical, developmental/high potential, strategic/executive, functional/ tactical), while as Newton et al. (2007) stressed upon the substantial differences between the functions of international transfers (position filling, management development, organizational development). However the fact is that the majority of the reentry literature approaches corporate reentry in a uniform manner. Consequently, this review also does not differentiate among different types of corporate returnees.

Inflow and outflow of expatriates is acknowledged to meet the need of skilled and professional manpower or human expertise in various fields (Harvey, 1989). In order for companies to stay competitive in the global market, it seems as though they are forced to send employees on International Assignments (Paik et al., 2002). Unfortunately, organizations seem to focus mainly on the expatriation process and more or less ignore the repatriation process (Harvey, 1982). Most people believe that coming back home will not cause any problems, when it in fact can be more problematic than settling into the new host country (Paik et al., 2002). The final strategy to effectively manage an International Assignment involves a deliberate repatriation process (MacDonald & Arthur, 2005; Paik et al., 2002). This process would require the skills and knowledge acquired on assignment in conjunction with positions available in the home location to be appropriately noted in order to assist with future career planning. The absence of such measures results in escalated cost to the employees, which comprises of disappointment with international experience, disappointment with the job level in the home office upon return from an International Assignment and failure to utilize newly acquired management skills, besides the feeling of career displacement or stagnation. The costs to the company include lost investments because of the underutilization of employees and the cost of their replacement when they decide to leave the firm. This supports the research oriented aspect of looking at the degree of perceived differences between host and home country cultures and adjustment scenarios, leading to improved adjustment (Jenkins & Mockaitiss, 2010).

Research shows that many Repatriates experience a number of work-related challenges upon return (Gomez-Mejia & Balkin, 1987; Suutari & Brewster, 2003). Repatriates report a lack of suitable career prospects available to them upon their reentry, a lack of challenging work and responsibilities, as well as a lack of understanding concerning their new experiences and knowledge on the part of their home-organization colleagues. This may result in individual career paths often being neglected in the strategic planning of International Assignments, culminating in the perception that international transition is a risky move for long-term career prospects of the repatriate (Naumann, 1992). Another work-related concern of returning employees is the ambiguity between the Repatriates and their HR managers, with respect to the definition of a successful repatriation (Paik et al., 2002). Returning individuals may like to focus on country and culture re-adjustment, while as the HR managers would primarily be interested in work-related outcomes and thus would expect the returning individuals to perform efficiently. Research demonstrates that multinational corporations seem to largely neglect the issue of reentry. Black et al. (1992) found among researched organizations, as few as 4.3% of U.S. organizations informed their employees of their return date with notice of half a year or more. Among the other issues reported as problematic are; the loss of autonomy, locally bounded technical and managerial obsolescence, a reduced range of responsibilities and the loss of within-company networks (McDonald, 1993). Caligiuri and Lazarova (2001) touched upon some considerations to be taken into account by managers attempting to create strategic repatriation policies. They stressed on the importance of coherence between selection procedures, expatriate performance management and the repatriation system and emphasized that, while substantial research and writings detail the retention strategies of the home-organization, a certain level of turnover upon repatriation is in fact functional and should be included in the strategic planning of International Assignments.

3.6.2 Indian Information Technology Sector: Nature of Growth and Implications

Information Technology is meant to act as a nodal institution for the promotion of the sector, facilitating and coordinating the various initiatives of the central and state governments and the private sector (Mathur, 2006). The early nineties initiated the golden era for information Technology in India. Liberalization, privatization and globalization, paved way for huge leaps in Indian IT Sector. Presently with total revenues of about $88.1 billion, Indian IT sector is one of strongest industries worldwide has grown for more than 30% for more than 20 years. Indian IT firms have moved up the value chain from low cost programming in the early years to providing premier global IT consulting services.

Prior to the global recession, in 2008, the IT industry flourished due to the exponential increase in global IT spending. However recession also created a pent-up demand for IT. Being a service oriented industry, this sector is highly dependent on manpower and due to the expansion in IT sector in terms of form, volume etc., there is a continuous demand for the demographic dividend, in form of Engineering graduates, skilled workforce etc.

Some of the challenges faced by Indian IT Industry are; uncertain global economy, protectionist measures of source and destination nations, tough completion from MNC's, pressure on billing rates, low employee utilization, volatility of Indian rupee against the dollar. Despite the economic upheaval, Indian IT firms are coping up with the various challenges, by innovating, finding new markets, and new customers and by developing new products etc. Wipro has started offering to IT solutions to the petroleum industry in Ghana; is leading to transformations. The use of cloud computing and mobile devices is slowly increasing. Infosys has added about 20 new clients in the past quarter in cloud, big data and security, (Satyamoorthy, 2006) (Fig. 3.3).

With the increase in use of apps on mobiles, programs are being written for mobile. Also companies have started positioning themselves as IT consulting firms. In the near future, government of India might provide the impetus as it plans to go big on digitization, as is evident by the Prime Ministers "DIGITAL INDIA" and "MAKE IN INDIA" campaign. For these initiatives there is an opportunity for further growth wherein manpower expertise would be required for skilled partnership managers, well versed in areas of digital technologies, big data analytics, mobile application development, cloud services and cyber security services (NASSCOM, 2015).

Given the relevance of the aforesaid context, it is evident that nearly similar precursors and outcomes may be applicable in the Indian Corporate perspective and therefore requires similar deliberations and analysis, given the fact that post 1991 the

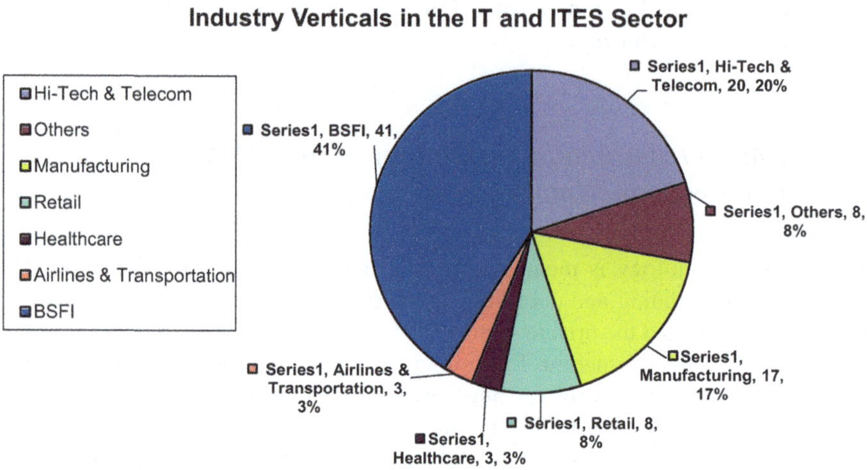

Fig. 3.3 Industry verticals in the IT-sector. *Source* NASSCOM & iMaCS Analysis (2015)

Indian government has introduced several reforms and policies to make India a global economic power, which has paved way for various alterations made to open up the system, attract foreign investors, send home country nationals abroad on assignments and to decrease the bureaucratic hassles in implementation of such policies. India's software and services exports have been rising rapidly over the last decade, thereby leading to gains from trade. Indian firms provide essential maintenance and development services, enabling overseas firms to use their scarce in-house IT staff for higher value added work, such as design and develop new types of applications. The growth in this sector is driven by steady increase in scale and depth of existing service lines; addition of newer vertical-specific and emerging niche business services, continued expansion of service portfolios and higher value processes. The IT industry is highly export oriented. Some Indian IT companies have already moved into significantly higher value added segments such as mission- critical applications, development and support, product design, HR management, knowledge process outsourcing for pharmaceutical companies and large complex projects. This calls for a holistic analysis of the corporate repatriation scenario in Indian context after an in-depth analysis of the relevant literature.

3.7 Scope of the Study

The two dimensions of International Assignment are Expatriation and Repatriation. Outcome of studies in India which explore the dimensions associated with Repatriation can enhance the prospects of minimizing the problems associated herein and thus maximizing the potential advantages an International Assignment offers. From the organizational perspective, the reasons why Repatriated employees are valuable assets are:

- As expatriates, employees can act as mentors to future expatriates by imparting skills like transfer of technology, managing cross-culturally, negotiating in a specific culture, communicating effectively in a host culture.
- They can bring forth a fresh approach to information gathering, analysis of data, and problem-solving as a result of having work cross-culturally in an effective manner.
- They may have become more flexible in changing circumstances. Due to the fact that different approaches have been tried in other contexts, they may be able to bring insights and innovation to the planning process.
- The returned expatriate who has performed at a high level in a host culture may bring a dimension of confidence and competence that will enhance his or her value to the company as it competes in a changing world market.
- Having worked outside the culture of the company and the country, the Repatriated employee may well have insights that can effect needed change.
- The Repatriated employee will likely bring a greater tolerance for divergent ideas and less structured and orderly contexts, in which to work.

The corporate goal to secure an acceptable return on investment while implementing the company's global strategy warrants that the company retains the Repatriates and utilizes the experience and knowledge that they have gained. Reduction in attrition costs associated with unsuccessful repatriation after International Assignments would optimize the high training costs for replacements and minimize the loss of return on investment in salaries and benefits paid to the employees while on the assignment (Black & Gregersen, 1991a; 1991b). Such futuristic goals have prompted management and human resources professionals to consider various options for reducing the cost of the International Assignments, such as altering the structure and terms of assignments and revising their repatriation policy and practice. While there is still much to be learned about Repatriates' turnover intention and its antecedents and outcomes, this study has made an attempt to identify the factors which contribute for Repatriates Adjustment. First, its results consolidate previous literature, by providing empirical evidence that Repatriation Adjustment and duration of the International Assignment are positively related. Repatriation Adjustment was the strongest predictor of intent to leave the organization for Indian MNC's Repatriates after repatriation. The Repatriates who perceived a higher level of Repatriation Adjustment had a lower intent to leave.

Second, this study is the first research that identifies Repatriates adjustment among Indian MNC Repatriates. Therefore, the conceptual framework of this study can be a guide to future research in Repatriates' adjustment. With the globalization of economies, the large amount of capital invested in international personnel, the repatriation process requires further attention. This study provides evidence about the difficulty of repatriation process and reports that cultural re-entry intention to leave and career development programs, duration of the International Assignment were the major factors influencing Repatriates' adjustment. A research oriented approach towards the discussion of repatriation is called for, in order to address the issues related to the process underlining some conditions for successful expatriation namely providing pre departure training and setting of realistic goals, addressing the culture shock phenomenon, provide training related to the process of reentry, address financial issues, offer a realistic job preview, bring motivations in line with company perspective, provide mentors and career planning along with optimum utilization of Repatriate talent.

A vast database of IT employees who find themselves as expatriates in foreign set ups and after a professional stint, return back as Repatriates, going through the cycle of expatriation and repatriation is noteworthy in this sector due to the outsourcing of business and also due to the fact that post liberalization India has emerged as a big player on the IT scene. Therefore tapping of the Repatriate talent is crucial in terms of the investment associated with it and also due to the fact that competency acquired doesn't go waste. Therefore the research area as implied here covers the IT sector in north India. The employees who comprise the sample were taken from the firms operating in this sector. The purview of the present research was to analyze the manifold dimensions of repatriation as it exists in the modern day IT Corporate set up in India. Keeping in view the justification and objectives of conducting the present research, data was collected from Repatriates of various IT-Companies e.g. Sapient Technologies, HCL, Informatica, Accenture, Aricent, Xerox, I-Gate, Amdocs, Infosys, Tech Mahindra, TCS, Wipro, Hexaware etc.

3.8 Justification for the Study

Companies need employees that are willing to work globally, since the most effective way to achieve international experience is by living and working within a foreign business arena (Engen, 1995). Effective use of Repatriated employees can provide a source of experience and innovative thinking strategies to enhance the competitive position of organizations. The cost of unsuccessful ventures can result in failure of a firm on financial, strategic and other associated parameters of successful expatriation and repatriation. Repatriates are an invaluable repository of knowledge and thus they have an irreplaceable role in organizational learning, by virtue of transferring that knowledge from host countries to headquarters, and vice versa.

The high cost of failed International Assignments impacts the financial aspects of the organization and an employee, in an adverse manner (Webb, 1996). Given this strategic fallout, the ways to predict Repatriate retention and lower their turnover is an important challenge facing organizations today (Black et al., 1992, 1995; Stroh & Gregerson, 1995). Human resource departments, in particular, have great responsibilities to ensure a smooth re-adjustment for employees returning to home country.

From time to time various research based theories have been propounded to analyze and validate the importance of successful repatriation. Martin and Harrell, (1996) in their Expectancy value model suggested that the assigned employees have expectations about their eventual return to the home office. Taking this forward, the Expectancy Violation Theory (Mooradian, 2004) states that unfulfilled expectations can be violated either positively or negatively. The levels of repatriation general adjustment, from highest to lowest, occurred with over met expectations, met expectations, and under-met expectations (Black, 1991). Reentry systems theory, according to Martin and Harrell (1996), concentrates on the importance of communication and contact with the employees before, during and after the assignment. Similarly the Equity theory of repatriation focuses on the employee motivation and commitment dependent on company policy or irrespective which is a deciding factor in determining the retention of the employee. The "W curve theory" assimilates the combined U curves; one of Expatriation and the other of Repatriation and the associated four stages of adjustment namely the excitement or the honeymoon stage on return and arrival, the hostile and emotional stage, the identity crisis stage and recovery stage and the final adjustment stage. Facilitation of these stages is a necessity to transfer the acquired competencies of Repatriate and acculturate him. This can be achieved by the tools of pre departure training, addressing the Reverse Culture Shock, providing realistic job preview, keep the perspectives of employee and organization in line and strive for effective career planning. All these parameters have to be addressed in order to retain, harness and accelerate the benefits of repatriation. Despite a plethora of studies done within this field, there is a dearth of guidelines on how to handle repatriation in an effective way (Goss & Hynes, 2005).

Taking into account the manifestations of the context and findings of Repatriation theories and consequent research, it becomes evident that there exists a research gap in the domain of Repatriate adjustment in India and its implications on various social and organizational criterions. Indian expatriates are in increasing demand by Indian enterprises which are growing into regional and global competitors. The main focus of this study is to analyze the dimension of Indian Repatriate acculturation in multi faceted dimensions of the factors taken into consideration and its analysis in the Information Technology Industry. Unlike before, the context of Competency Transfer, Anticipatory Adjustment, Effective Repatriation and Reverse Culture Shock has been delved upon, in a single empirical study on Repatriation scenario and its associated nuances in Indian Information technology Sector.

An overall cycle starting with expatriation to an analytical expatriation monitoring and proceeding towards successful repatriation, thus reaping the benefits of the human capital developed in the process, is the consequent prerogative and endeavor of any corporate enterprise. A comprehensive study to help improve the chances of finding the most appropriate actions, meeting the demands of expatriates, and create a smooth repatriation process has been attempted by the present study, keeping in view the afore mentioned aspects of Repatriate Anticipatory Adjustment, Competency Transfer, Effective Repatriation and the effect of Reverse Culture Shock on Indian IT-Sector Repatriate adjustment.

References

Adler, N. J. (1981). Re-entry: Managing cross-cultural transitions. *Group and Organizational Studies, 6*(3), 341–356.

Adler, N. J., & Gunderson. (2008). *International dimensions of organizational behavior*. Thomson South Western.

Allen, D., & Alvarez, S. (1998). Empowering expatriates and organizations to improve repatriation effectiveness. *Human Resource Planning, 21*(4), 29–39.

Ashford, S. J., & Taylor, M. S. (1990). Adaptation to work transitions: An integrative approach. *Research in personnel and human resources.*

Baloch, M. T. (2013). *The effect of repatriation on managers returning from foreign assignments.* Department of International Business, Dublin Business School, Dublin, Ireland.

Bandura, A. (1991). Social cognitive theory of self regulation. *Organizational Behavior and Human Decision Processes, 50*(2), 248–287.

Baruch, Y., & Altman, Y. (2002). Expatriation and repatriation in MNCs: A taxonomy. *Human Resource Management, 41*(2), 239–259.

Befus, C. P. (1988). A multi-level treatment approach for culture shock experienced by sojourners. *International Journal of Intercultural Relations, 12*(4), 381–400.

Bell, N. E & Staw, B. M. (1989). People as sculptors versus sculpture: the roles of personality and personal control in organizations. In *Handbook of career theory*. books.google.com.

Black, J. S., & Gregersen, H. B. (1991a). When Yankee comes home: Factors related to expatriate and spouse repatriation adjustment. *Journal of International Business Studies, 22*(4), 671–694.

Black, J. S., & Gregersen, H. B. (1991b). The other half of the picture: Antecedents of spouse cross-cultural adjustment. *Journal of International Business Studies, 22*(3), 461–477.

Black, J. S., Gregersen, H. B., & Mendenhall, M. E. (1992). Toward a theoretical framework of repatriation adjustment. *Journal of International Business Studies, 23*(4), 737–760.

Black, J. S., Mendenhall, M. E., & Oddou, G. (1991). Toward a comprehensive model of international adjustment: An integration of multiple theoretical perspectives. *Academy of Management Review, 16*(2), 291–317.

Bossard, A. B., & Peterson, A. B. (2005). The repatriate experience as seen by American expatriates. *Journal of World Business, 40*(1), 9–28.

Brett, J. M. (1983). Coping with new jobs: A comparative study of new hires and job changers. *Academy of Management Journal., 46*(7), 233–267.

Brewster, C., Sparrow, P., Vernon, G., & Houldsworth, E. (2007). *International human resource management*. cipd.co.uk.

Caligiuri, P., Phillips, J., Lazarova, M., Tarique, I., & Burgi, P. (2001). The theory of met expectations applied to expatriate adjustment: The role of cross-cultural training. *The International Journal of Human Resource Management, 12*(3), 357–372.

Chamove, A. S., & Soeterik, S. M. (2006). Grief in returning sojourners. *Journal of Social Sciences, 13*(3), 215–220.

Church, A. T. (1982). Sojourner adjustment. *Psychological Bulletin, 91*(3), 540–572.

Dowling, P., Festing, M., & Engle, E. D. (2008). *International human resource management: Managing people in a multinational context*. books.google.com.

Engen, J. R. (1995). Coming home. *Training, 32*(3), 37–40.

Feldman, D., & Tompson, H. (1993). Expatriation, repatriation, and domestic geographical relocation: An empirical investigation of adjustment to new job assignments. *Journal of International Business Studies, 24*(3), 507–529.

Fenwick, M. (2004). On international assignment: Is expatriation the only way to go? *Asia Pacific Journal of Human Resources, 42*(3), 365–377.

Friedman, M. (2007). *The social responsibility of business is to increase its profits* (pp. 173–178). Springer.

Furuya, N., Stevens, M. J., Oddou, G., Bird, A., & Mendenhall, M. E. (2006). The effects of HR policies and repatriate self-adjustment on global competency transfer. *Asia Pacific Journal of Human Resources, 45*(1), 6–23.

Gaw, K. F. (2000). Reverse culture shock in students returning from overseas. *International Journal of Intercultural Relations, 24*(1), 83–104.

Glisson, C., & Durick, M. (1988). Predictors of job satisfaction and organizational commitment in human service organizations. *Administrative Science Quarterly, 33*(1), 61–81.

Gomez-Mejia, L., & Balkin, D. B. (1987). The determinants of managerial satisfaction with the expatriation and repatriation process. *Journal of Management, 14*(5), 213–220.

Goss, Y., & Hynes, G. (2005). Strategies for successful repatriation. *Journal of International Business Research, 4*(1).

Gullahorn, J. T., & Gullahorn, J. E. (1963). An extension of the U-curve hypothesis. *Journal of Social Issues, 19*(3), 33–47.

Harvey, M. G. (1989). Repatriation of corporate executives: An empirical study. *Journal of International Business Studies, 20*(1), 131–144.

Howe-Walsh, L. (2010). *The road to repatriation: Implications for HR policy and practice* [Ph.D thesis]. University of Portsmouth.

Hyder, A., & Lovblad, M. (2007). The repatriation process—A realistic approach. *Career Development International, 12*(3), 264–281.

Jenkins, E. M., & Mockaitiss, A. I. (2010). You're from where? The influence of distance factors on New Zealand expatriates' cross-cultural adjustment. *The International Journal of Human Resource Management, 21*(15), 2694–2715.

Kraimer, M. L., & Wayne, S. J. (2004). An examination of perceived organizational support as a multidimensional construct in the context of an expatriate assignment. *Journal of Management, 7*(3), 213–226.

Lazarova, M., & Caligiuri, P. M. (2001). Retaining repatriates: The role of organizational support practices. *Journal of World Business, 36*(4), 389–401.

Lazarova, M., & Cerdin, J. (2007). Revisiting repatriation concerns: Organizational support versus career and contextual influences. *Journal of International Business Studies, 38*(3), 404–429.

Lazarova, M., & Tarique. (2005). Knowledge transfer upon repatriation. *Journal of World business*, 40(4), 361–373.

Linehan, M., & Scullion, H. (2002). Repatriation of European female corporate executives: An empirical study. *The International Journal of Human Resource Management, 13*(2).

Macdonald, S., & Arthur, N. (2005). Connecting career management to repatriate adjustment. *Career Development International, 10*(2), 145–158.

Martin, J. N., & Harrell, T. (1996). Reentry training for intercultural sojourners. In *Handbook of intercultural training* (pp. 307–326). Sage Thousand Oaks.

Mathur, S. K. (2006). Indian information technology industry: Past, present and future and a tool for national development. *Applied Information Technology.*

Maybarduk, S. M. (2008). An exploration of factors associated with reentry adjustment of US Foreign Service spouses. dspace.smith.edu

Mendenhall, M. E. (2007). Elusive yet critical challenge of developing global leaders. *European Management Journal, 24*(6), 422–429.

Mendenhall, M., & Odou, G. (1985). Dimensions of expatriate acculturation: A review. *Academy of Management Review, 10*(1), 39–47.

Mooradian, B. L. (2004). Going home when home does not feel like home: Reentry, expectancy violation theory, self construal and psychological and social support. *Intercultural Communication Studies, 24*(5), 123–140.

NASSCOM. (2015, October). Job growth in IT to slow down. *Times of India, Times Business*, 6 October, 2015.

Naumann, E. (1993). Antecedents and consequences of satisfaction and commitment among expatriate managers. *Group and Organization Management.*

Newton, S., Hutchings, K., & Kabanoff, B. (2007). Repatriation in Australian organizations: Effects of function and value of international assignment on program scope. *Asia Pacific Journal of Human Resources, 45*(3), 295–331.

Oberg, K. (1960). Cultural shock: Adjustment to new cultural environments-curare,—agem-ethnomedizin.de

Paik, Y., Segaud, B., & Malinowski, C. (2002). How to improve repatriation management: Are motivations and expectations congruent between the company and expatriates? *International Journal of Manpower, 23*(7), 635–648.

Reiche, S. B. (2012). Knowledge benefits of social capital upon repatriation: A longitudinal study of international assignees. *Journal of Management Studies, 49*(6), 1052–1077.

Satyamoorthy, S. (2006). http://www.mbaskool.com/business-articles/operations/4935-indian-it-sector-an-analysis.html

Selmer, J., Chiu, R. K., & Shenkar, O. (2007). Cultural distance asymmetry in expatriate adjustment. *Cross Cultural Management: An International Journal, 14*(2), 150–160.

Shrinivas, P. B., Harrison, D. A., Shaffer, M. A., Dora M. L. (2005). Input-based and time-based models of international adjustment: Meta-analytic evidence and theoretical extensions. *Academy Management Journal, 48*(2), 257–281.

Solomon, E. (2001). The dynamics of corporate change: Management's evaluation of stakeholder characteristics. *Human Systems Management, 20*(3), 257–265.

Sparrow, P. R., & Braun, W. (2007). *HR sourcing and shoring: Strategies, drivers, success factors and implications for HR*. Lancaster University.

Stroh, L., & Caligiuri, P. M. (1997). *Increasing global competitiveness through effective people management*. The Global Leadership Institute.

Stroh, L. K., & Gregerson, H. B. (1995). Coming home to the Arctic cold: Antecedents to Finnish expatriates and spouse repatriation adjustment. *Personnel Psychology, 50*(3), 635–654.

Stroh, L. K., Gregersen, H. B., & Black, J. S. (1998). Closing the gap: Expectations versus reality among repatriates. *Journal of World Business, 33*(2), 111–124.

Sundquist, J., & Johansson, S. E. (1996). The influence of exile and repatriation on mental and physical health: A population-based study. *Social Psychiatry and Psychiatric Epidemiology, 31*(1), 21–28.

Sussman, N. M. (2000). The dynamic nature of cultural identity throughout cultural transitions: Why home is not so sweet. *Personnel and Social Psychology Review, 4*(4), 335–373.

Sussman, N. M. (2001). Repatriation transitions: Psychological preparedness, cultural identity, and attributions among American managers. *International Journal of Intercultural Relations, 25*(2), 109–123.

Suutari, V., & Brewster, C. (2003). Repatriation: Empirical evidence from a longitudinal study of careers and expectations among Finnish expatriates. *International Journal of Human Resource Management, 14*(7), 1132–1151.

Suutari, V., & Valimaa, K. (2002). Antecedents of repatriation adjustment: New evidence from Finnish repatriates. *International Journal of Manpower, 23*(7), 617–634.

Szkudlarek, B. S. (2010). *Reentry—A review of the literature*. Department of Organization and Personnel Management, Rotterdam School of Management, Erasmus University Rotterdam, Rotterdam, Netherlands.

Tahir, A. H. M., & Ismail, M. (2007). Cross-cultural challenges and adjustments of expatriates: A case study in Malaysia. *Alternatives: Turkish Journal of International Relations, 6*(3, 4).

Taveggia, C. T., & Santos, N. G. L. (2001). Cross cultural adjustment: A test of the uncertainty reduction principle. *International Journal of Cross Cultural Management, 1*(2), 153–171.

Tharenou, P. (2002). Receptivity to careers in international work—Abroad and at home. *Australian Journal of Management, 27*(1), 129–136.

Thompson, C. L., & Christofi, V. (2006). Can you go home again? A phenomenological investigation of Cypriot students returning home after studying abroad. *International Journal for the Advancement of Counseling, 28*(1), 21–39.

Tivey, T. L. (2008). *Reflections on the efficacy of the human resource management strategies of international organizations regarding expatriation, repatriation and relocation* [M.Sc dissertation]. University of Nottingham.

Torbiorn, I. (1982). *Living abroad: Personal adjustment and personnel policy in the overseas setting.* John Wiley & Sons.

Tung, R. L. (1998). American expatriates abroad: From neophytes to cosmopolitans. *Journal of World Business.*

Vidal, M., Valle, R., Aragón, M. I. B., & Brewster, C. (2007). Repatriation adjustment process of business employees: Evidence from Spanish workers. *International Journal of Intercultural Relations, 31*(3), 317–337.

Webb, A. (1996). The expatriate experience: Implications for career success. *Career Development International, 1*(5), 38–44.

Welch, D. E., Lawrence, S., & Worm, V. (2007). The international business traveller: A neglected but strategic human resource. *The International Journal of Human Resource Management, 18*(2), 173–183.

White, D., Absher, K., & Huggins, K. (2011). The effects of hardiness and cultural distance on sociocultural adaptation in an expatriate sales manager population. *Journal of Personal Selling and Sales Management, 31*(3), 325–337.

www.ibef.org/industry/indian-iT-and-iTeS-industry-analysis-presentation

Yan, A., Zhu, G., & Hall, T. (2002). International assignments for career building: A model of agency relationships and psychological contracts. *The Academy of Management Review, 27*(3), 373–391.

Chapter 4
Research Design and Methodology

4.1 Introduction

Keeping in view the objectives of the present research study, the process and methodology used for carrying it out has been elaborated in this chapter. The methodology adopted for this study incorporates a number of activities directly or indirectly used in conducting the research related process and the subsequent collection of data. The research process of the study began with the selection of the research problem for the formulation of the basic premise and scope, besides fathoming the possible limitations of the study.

The logical step in the research process after the formulation of the research objectives, identification of the research variables and development of the theoretical framework, was to design it in such a manner that the research data collected and analyzed is in conformity with the laid down objectives. The rationale for determining the universe of sample and the construction of the Questionnaire is determined. For data analysis and its interpretation, techniques and statistical methods that have been employed are elaborated upon in the chapter. The research methodology used for the collection and interpretation of the research data is essentially based on the paradigms related to Repatriation Adjustment and management.

4.2 Research Purpose and Scope

The present study intends to examine the antecedents of Repatriation Adjustment in the corporate sector India, with significant emphasis on the variables of Reverse Culture Shock, Anticipatory Repatriate Adjustment, Competency Transfer and Effective Repatriation. The study investigates the inter-linkages of the four mentioned constructs in the context of Repatriation Adjustment and how this results in associated Job, Organizational, Individual and other significant implications. Research in

M. B. Nag and F. Ahmad Malik, *Repatriation Management and Competency Transfer in a Culturally Dynamic World*,
https://doi.org/10.1007/978-981-19-7350-5_4

the domain of Repatriation is not as exhaustive as it ought to be, given the dynamics of a globalized world. Despite prior studies done in this field, there are very few recommendations or adequate guidelines, on the best way to handle Repatriation in Indian IT sector, which exhibits a widespread phenomenon of Expatriation and Repatriation of employees, due to a liberalization and global business scenario in the recent years. In the context of Information Technology (IT) services, Indian software firms provide IT solutions to organizations all around the globe.

On statistical analysis of variable responses of Effective Repatriation, Anticipatory Repatriation Adjustment and Reverse Culture Shock, via a questionnaire, it was observed that interplay of the said constructs resulted in transfer or failure of transfer of competencies of repatriates or ineffective repatriate adjustment. The demographics at play while collecting data and measuring the reliability and validity of constructs, indicated the respective outcomes. Given the fact that the current study validates the above given benchmarks, it is imperative that International HR practices be designed so as to enable Repatriate satisfaction, which in turn would manifest into the firm's success in the global arena.

International assignments are the pivot of dynamism and concern to many corporations who make big investments in overseas assignments. As more companies expand their business overseas, the need to send employees on international assignments and retaining them has increased, making the process of discovering the reasons for these failures increasingly thus becomes important. Thus, the main aim of this paper is to explore the holistic perspective of Repatriation after International Assignments.

4.3 Research Objectives

On the basis of the literature reviewed, hypotheses and the domain of the study, the following objectives of the research were set:

Objective 1: To study and evaluate the emerging phenomenon of Repatriation Management, in the context of IHRM

The first objective intended to study the current expansion and internationalization of business organizations, wherein organizations are rapidly moving from being purely domestic players to being cross-national players. Organizational preference for employees with global skills has increased and hence there is an increase in expatriation and consequently the need for Effective Repatriation. Human resource departments, specifically, have incredible responsibility to guarantee a smooth re-entry for employees coming back to home country. While several researchers have examined expatriation, few have studied the last part of international assignments—the Repatriation Process. The essential purpose behind repatriation not being seen as an important aspect is due to the belief that the expatriate is coming home and therefore it should not result in any issues. On the other hand, studies have demonstrated that repatriating is frequently more difficult than expatriating, and that there are both practical as well as psychological problems, related to the Repatriation Process. The cost of failed international assignments is high, both financially for the organization and from an employee's perspective as underutilization of employee talent leads to attrition. Organizations should precisely consider the global expertise and knowledge that the repatriate acquires and identify their definite value, since the cost of an International assignment is very high.

Existing information on global human resource practices recommends that customary parameters of human resource management demonstrate inadequate management of portability needs of managers and also does not acknowledge their choices, with respect to international transfers. Consequently, guidelines and more proposals on how to successfully manage repatriation are required. Repatriation in Corporate India within the domains of the constructs described in the study, namely Anticipatory Adjustment, Competency Transfer, Effect of Reverse Culture Shock and subsequently the credibility of Effective Repatriation, are intended to have a deeper insight of the process. The findings of this research can be deliberated upon and analyzed to provide a base for a discussion about how and what practical solutions that companies could apply to meet the underlying needs in creation of an Effective Repatriation process.

A statistical analysis validated this objective by collection of data from NASSCOM accredited IT companies and subjecting it to various statistical tools, which proved that Repatriation concerns do effect the Expatriates returning back to home country as Repatriates. The results indicated that increase in the effectiveness of Repatriation process is significantly predictive of overall Repatriate and organizational welfare. The summation of the findings of the two hypotheses, using statistical tools like Correlation and Multiple Regression analysis, confirm the said objective.

Objective 2: To analyze the factors effecting Repatriate Adjustment in the home country

One of the essential theoretical procedures identified with change, whether it is repatriation adjustment, cross cultural adjustment, or adjustment after a domestic transfer, is the reduction of uncertainty (Black et al., 1999a; Brislin, 1981). Individuals enter new environments and many a times maybe unclear with in distinguishing between adequate and unsuitable, or proper and wrong. The adjustment procedure includes this learning, in order to lessen the ambiguity between such choices and subsequent practices (Brislin, 1981). The concept of "cultural shock" is generally an arrangement of negative responses of people to experiences in an alien atmosphere which reflects that they do not have a complete and exact arrangement of schematas for comprehension and assimilation in the new, circumstance (Torbiorn, 1982).

The objective is to investigate the factors effecting Repatriate Adjustment in the home country. The four constructs with their respective items under study analyzed the said perspective in detail wherein the transfer of competencies as a result of Effective Repatriation techniques aided the adjustment antecedents of a Repatriate. Likewise, the negative impact of Reverse Culture Shock in the home country on repatriation, which may cause impediments in the fulfillment of Anticipatory Adjustment expectations while abroad, has been analyzed in detail, thereby holistically covering the process of Repatriate adjustment back in the home country. It has been empirically analyzed in the study pertaining to the first hypothesis that Reverse Culture Shock in the home country does impact Anticipatory Adjustment expectations of Repatriates wherein, home-country organizations lose out in form of unsuccessful repatriation process involving costs for both the organization and the employee. The study looks at the factors influencing Repatriation adjustment and satisfaction from the perspectives of Indian IT Sector.

Objective 3: To examine the relation between Repatriation and Reverse Cultural Shock

The study elucidated research literature from various sources and elaborates the psychological, individual, sociological and organizational impediments that go into giving shape to the experience of Reverse Culture Shock during home country or in-country repatriation adjustment. This manifests itself in the context of adjusting to work, culture and general environment, wherein the antecedent variables comprise Individual, Family, Job, Organizational and non work adjustment (Black et al., 1992).Thus, the problem identified and analyzed is that of the impact of Reverse Culture Shock and repatriates' difficulty in adjusting after an international assignment, leading to possible attrition and wastage. The phenomenon of Reverse Culture Shock and its impact on such Anticipatory Adjustment has been statistically analyzed in the study.

Objective 4: To undertake an impact assessment of repatriation on Competency Transfer, in the context of home country organizations

The domain of Competency Transfer of Expatriates on their Repatriation is one of the primary focus areas of the study. This factor comprises issues pertaining to the transfer of competencies of Repatriated employees per se. Research suggests that if international assignments were designed keeping in mind the different aspects of skill and acumen transfer principles, as elucidated in the variables under the construct of Competency Transfer, Repatriate talent would get optimized. It signifies that Repatriates who are able to make use of their overseas acquired acumen, resulting from their international experience tend to feel more satisfied with their career growth. Consequently, it calls for policies that enhance expatriation, warrant superior performance, manage repatriation and lead to employee satisfaction.

Therefore the objective was to undertake an impact assessment of repatriation process on Competency Transfer in the context of home country organizations.

Objective 5: To propose industry specific strategic model for Effective Repatriation

The dynamics of International Human Resource Management functions with a time lapse which manifests in a gap between strategy implementation and practice as far as Repatriation management is concerned can lead to an unplanned and unstructured array of services and support mechanisms. The term effective is "successful or achieving the results that you want" (Cambridge Dictionary, 2005). Effective Repatriation is achieved when the repatriate acknowledges the effectiveness of repatriation process holistically, and when all parties involved, achieve the desirous results.

A successful repatriate brings home an assortment of valuable skills, knowledge, and attributes and creates opportunities to grow, both culturally and technically. Therefore this growth needs to be accommodated in adaptation processes undertaken on return to their home country, thereby providing a sense of career continuity and validating the value that the company places on expatriate assignments. The construction of an Effective Repatriation strategy is a key factor in retaining valued personnel and a prominent component of HRM process.

Various factors influence repatriation and in accounting for these variables, models for developing an Effective Repatriation support process have been proposed by various researchers, keeping in view the peculiar nuances of the socio-political and cultural environment of a country, besides the general and organizational specifics, as applicable. Jassawalla et al. (2004a, 2004b) proposed that the quality of interaction with sponsors and perception of support upon return were found to be relevant. Reverse Culture Shock, repatriation adjustment, managing expectations, welcome home orientation, valuing international experience were also found to impact repatriation adjustment (Martin & Anthony, 2006). Black gave an integrated model of International Adjustment which had three dimensions: Work Adjustment, Interaction Adjustment with foreign country nationals, Cultural Adjustment. The factors affecting expatriate adjustment are (Black & Gregersen, 1991): Anticipatory Adjustment factors: Individual (pre-departure training, previous international experience etc.) and Organizational (effectiveness of selection procedure); In-country Adjustment factors: post departure adjustment issues.

As per the extensive analysis of Repatriation literature and that pertaining to the constructs of the present study and basis the findings of the statistical analysis of the data collected, in line with the proposed hypothesis, a specific strategic model for Effective Repatriation is proposed, wherein precisely a framework of four broad factors have an impact on Repatriation adjustment, from Organizational point of view, in eliminating the bottlenecks associated with Repatriation. These are; Policy development, which caters to recognizing the possible stress, repatriation may cause and the support of the company during time of transition.

The respective analysis of the constructs which have lead to conclusive findings as deliberated upon earlier are that an Effective Repatriation strategy should primarily comprise of integrated components, dealing with the principles and essence of the repatriation strategy, translated into a formal documented agreement, enacted through a supportive program. A viable model of Effective Repatriation based on data from a study of repatriation cases from Indian IT industry has been proposed which caters to the feasibility criterion of the items predominantly at play, in making of the said constructs of the study. This study is also novel because in-depth studies of Repatriation adjustment and Effective Repatriation on Indian corporate sector are scarce and industry practitioners will always need to improve the retention of their international human resource through productive repatriation outcomes from organizational and employee perspective.

4.4 Resarch Hypotheses

H1: Reverse Culture Shock impacts Anticipatory Repatriation Adjustment

The hypothesis corresponds with the second objective of the present study, wherein correlation, exploratory factor analysis and multiple regression analysis were conducted on the collected data; to test whether Reverse Culture Shock impacts Repatriation adjustment.

This implies that the findings considerably denote that the criterion of the variables defining Anticipatory Adjustment, rely heavily on the construct of Reverse Culture Shock and that the latter has an impact on the former whereby the Anticipatory Adjustment while overseas with respect to the foreseen adjustment in home country, gets negatively impacted. Therefore validating the Hypothesis that Reverse Culture Shock impacts Anticipatory Adjustment of Repatriates.

H2: Competency Transfer is significantly a function of Effective Repatriation

The hypothesis corresponds with the second objective of the present study, wherein exploratory factor analysis to arrive at factor groups and multiple regression analysis was performed in order to test whether Effective Repatriation is a significant predictor of Competency Transfer of a repatriate. Studies have reported that expatriates receive significant personal and self-development benefits from foreign assignments. They gain a broader, global perspective, skills, expertise and acumen on the firm's operations and these types of development can have positive career and organizational implications. It is desired that HR managers devise practices to better manage this critical resource (the expatriates) in order to maximize the wealth of knowledge and experience gained from international assignments. Specifically, it has been analyzed that the repatriates value more positively the extent to which their jobs have task variety and autonomy, provide opportunities for learning, allow them to apply their knowledge and give them enough responsibility.

Through an empirical approach, the dimensions that influence the role of repatriates in learning and sharing among headquarters and their subsidiaries, was examined. The findings suggest that the reasons that impact knowledge, acumen and Competency Transfer are connected to the muti-faceted dimensions of Effective Repatriation which were formulated as per previous research and its statements statistically correlated with Competency Transfer dimensions. In spite of the copious writing in information management, only a couple of observational studies have investigated knowledge management regarding international assignees. This phenomenon has a special relevance in the Indian context, since except for a couple of papers, there are no empirical studies in an exhaustive manner concerning this issue that involves Indian international assignees. The IHRM reality in Indian context is incipient as far as internationalization and knowledge management is concerned, and the organizational and national culture presents attributes that are different from the most profoundly investigated realms in countries other than India.

4.5 Methodological Framework

To achieve the objectives laid down and keeping in view the peculiarities for the study, a structured instrument in the form of a Questionnaire, was utilized for the study. All relevant and acknowledged measures of the constructs that are studied were derived from previous research.

4.5.1 Research Approach

Saunders et al. (2007) illustrated the difference between deductive approach in which we begin to develop the hypothesis and then design the research strategy to verify the validity of developed hypothesis and then design a research strategy to verify the validity of developed hypothesis and the inductive approach in which the collection of data comes first and then we analyze the data to develop the hypothesis.

In the present research we tried to test the hypotheses related to various independent variables and their relationship with the dependent variables to analyze the totality of Repatriation Adjustment in Indian IT Sector. The data was then collected and analyzed to test these hypotheses. Therefore the present research is deductive in nature. Deductive approach works from more general to more specific, where it is sometimes called the top down approach. Deductive approach begins with theory, hypothesis, observation and ends at confirmation.

4.5.2 Research Instrument (Questionnaire Development)

A Questionnaire was designed to measure the variables of the constructs under study. The objective was to study the impact of Reverse Culture Shock on Anticipatory Repatriation Adjustment and also to find out whether Competency Transfer is significantly a function of Effective Repatriation. A 5-point Likert scale (Strongly disagree, disagree, neither agree nor disagree, agree and strongly agree; was used to measure the values of the variables). The effects of the experimental treatments were measured on one dependent variable, Repatriation Adjustment which is measured in terms of whether or not, Effective Repatriation results in Competency Transfer, and whether Reverse Culture Shock widens the gap between anticipated and realistic expectations of repatriation.

I. Measures used in the study

The questionnaire included five sections, comprising of questions on Demographic profile of respondents, Competency Transfer, Anticipatory Adjustment, Effective Repatriation and Reverse Culture Shock.

Section A: Demographic Profile

This section deals with the respondents' demographic detail which is developed in order to gain information on significant variables such as, Gender, Age, Marital Status, Job Profile, Length Of Stay Overseas and Time Elapsed Since Return from Overseas Assignment.

Section B: Competency Transfer

This construct was measured with a 12-item scale (range 1–5) adopted by Furuya et al. (2007 and 2008), Löyvä (2014), Black et al. (1999b), Andersson (2006), Hermansson and Andersson (2008), Hermansson and Kilnes (2008), Santosh and Muthiah (2012). All items measuring Competency Transfer were ranked on a 5 point Likert scale (*1 = strongly disagree, 5 = strongly agree*).

Section C: Effective Repatriation

This construct was measured with a 13-item fit scale (range 1–5) adopted from Paganus (2005), Murray (2007), Santosh and Muthiah (2011) and Black and Gregersen (1991). All items measuring Effective Repatriation were ranked on a 5 point Likert scale (*1 = strongly disagree, 5 = strongly agree*).

Section D: Anticipatory Adjustment

This construct was measured on a 12-item scale adopted from Paganus (2005), Jennie et al. (2006), Santosh and Muthiah (2011). All items measuring Anticipatory Repatriation Adjustment were ranked on a 5 point Likert scale (*1 = strongly disagree, 5 = strongly agree*).

Section E: Reverse Culture Shock

This construct was measured 15-item. 5-point Likert scale, basis the scale fit from Masako, (2008), Niesen (2010) and Siddiqua (2010) papers on Reverse Culture Shock. The paper on Reverse Culture Shock amongst returning student sojourners by Gaw (2000) served as a reference.

4.5.3 Pilot Study

The testing of Questionnaire on a small sample of respondents from the specified study area was done. Improving the Questionnaire by identifying and eliminating potential problems, if any, was the purpose behind it. The motive of refining the Questionnaire provided a reality check on the clarity of the semantics involved in framing the statements and an analysis of the logic involved in the research, as planned. Pilot study tested and identified these nuances and as per the results, the wordings of three statements pertaining to Reverse Culture Shock construct, were rectified. Apart from this there were no other specific issues with respect to the Questionnaire design, wordings of statements or the procedure.

4.6 Research Design

A research design provides a framework for the collection and analysis of data. A choice of research design reflects the importance given to; expressing the causal connections between variables, generalizing to larger groups of individuals than those actually forming the part of the investigation, understanding the behavior and meaning of that behavior in its specific social context and having temporal (i.e. over time) appreciation of social phenomenon and their interconnections.

Survey technique was the research design opted for the current research. The survey is a method for collecting information or data as reported by individuals. Surveys are questionnaires (or a series of questions) that are administered to research participants who answer the questions themselves. Since the participants are providing the information, it is referred to as self-report data. Surveys are used to get an idea of how a group or population feels about certain parameters. When conducting a survey it is of the utmost importance that the participants answering the questions are a representative sample of the target population. In survey research, the researcher picks a sample of respondents from a population and provides them with a standardized questionnaire. The questionnaire, or survey, maybe in the form of a written document, which the respondent needs to fill, it can be an online questionnaire, a face-to-face interview, or a telephone interview.

4.6.1 Sampling Design

The sampling process is related to the right selection of right individuals, objects or events for the research study (Sekaran, 2006). This process consists of various steps which are taken by a researcher before actually starting the collection of data about the research problem.

4.6.2 Sample Population

Sample populace alludes to the whole set of individuals or things of interest that the researcher wishes to explore. According to Cooper et al. (2006), population is the total collection of sample estimates about which the inferences are drawn. As the primary objective of this study is to understand the Antecedents of Repatriation Adjustment in the Information Technology sector of India, the target population for this study consisted of Corporate Repatriates in this sector. After a thorough review of literature and preparation of a structured Questionnaire, the data was collected from Repatriate employees of NASSCOM accredited IT companies in India. A sample of 416 corporate executives participated in the study. Out of a total of 416, only 250 usable resources were found, resulting in a response rate of 60.1%. This satisfies the requirements of the research under scanner.

4.6.3 *Unit of Analysis*

The sample unit of analysis means the sample elements or the subjects of the study on which measurement is being undertaken (Kothari, 2004). A sample unit of analysis can be individuals, groups, a division, organization, an industry or countries. Hence, the selection of a particular unit of analysis, i.e. Repatriated employees of Indian Information Technology Sector, is based upon the nature, purpose and the research questions being addressed by the present study.

4.6.4 *Sampling Method*

The sample must be well representative of the population that it is selected from, so that the conclusions drawn from the sample are likely to be true to the population. Sampling method is the way of selecting the sample units of the study. Various sampling methods exist for use in research. Researchers distinguish between probability (random) and non-probability (non-random) sampling (Leedy & Ormrod, 2005). Probability sampling has the ability to ascertain that every element in the research population has an equal chance of being selected; thereby making the results representative.

Non-probability sampling is by nature non-representative and the probability of elements being selected cannot be guaranteed, nor can the extent to which generalizations may be limited to non- statistical arguments (Saunders et al., 2011). The nature of this study is more suited to non-probability sampling as the units of analysis need to adhere to the purpose and certain requirements specific to this research.

Therefore convenience random sampling strategy was used based on non-probability sampling design to select the domain from which the respondents were selected for the said research. In convenience sampling, the sample elements are selected purely on the basis of their availability.

4.6.5 *Sample Size Determination*

The sample size for the present study was selected using the following information:

1. **Error Estimation**

   Cochran's (1977) equation utilizes two key elements: (1) The risk the analyst is willing to acknowledge in the study, generally called the margin of error, or the error the researcher is willing to accept, and (2) the alpha level, the level of adequate risk the analyst is willing to acknowledge that the true margin of error surpasses the required margin of error; i.e., the probability that differences revealed by statistical analysis really don't exist; also known as Type I error.

Alpha level: The alpha level utilized as a part of deciding the sample size in most survey based research work is either 0.05 or 0.01 (Ary et al., 1996). In Cochrans' formulae, the alpha level is incorporated into the equation by using the t-value for the alpha level selected (e.g., t-value for alpha level of 0.05 is 1.96 for sample sizes above 120). Researchers ought to guarantee that they use the correct t-value (e.g., t-value for alpha of 0.05 and a population of 60 is 2.00. In general, an alpha level of 0.05 is acceptable for most research.
Acceptable margin of error: For categorical data, 5% margin of error is acceptable, and for continuous data, 3% margin of error is acceptable (Krejcie & Morgan, 1970). An increase in the values when a higher margin of error is acceptable or a decrease in the values when a higher degree of precision is required is permissible to the researcher.

2. **Variance Estimation**

   A critical component of sample size formulae is the estimation of variance in the primary variables of interest in the study. The researcher does not have direct control over variance and must incorporate variance estimates into research design. Cochran (1977) listed four ways of estimating population variances for sample size determinations: (1) takes the sample in two steps, and use the results of the first step to determine how many additional responses are needed to attain an appropriate sample size based on the variance observed in the first step data; (2) use pilot study results; (3) use data from previous studies of the same or a similar population; or (4) estimate or guess the structure of the population assisted by some logical mathematical results. A researcher needs to estimate the variance of scaled and categorical variables. In order to estimate the variance of a scaled variable, the comprehensive range of the scale, needs to be determined; thereafter divide the number of standard deviations that would include all possible values in the range, and then square this number. For example, if a researcher used a seven-point scale and given that six standard deviations (three to each side of the mean) would capture 98% of all responses, the calculations would be as follows:

$$S = 7(\text{number of points on the scale})/(\text{number of standard deviations})$$

4.6.6 Sample Size Formula

If we assume that a researcher has set the alpha level a priori at 0.05, plans to use a proportional variable, has set the level of acceptable error at 5%, and has estimated the standard deviation of the scale as 0.5., in that case, Cochran's sample size formula for categorical data and an example of its use is presented here along with explanation as to how these decisions were made.

$$n_\circ = (t)2 * (p)(q)/(d)^2$$
$$n_\circ = (1.96)2(0.5)(0.5)/(0.05)^2$$
$$n_\circ = 384$$

where,

- t value for selected alpha level of 0.025 in each tail = 1.96. (The alpha level of 0.05 indicates the level of risk the researcher is willing to take that true margin of error ay exceed the acceptable margin of error).
- (p) (q) estimate of variance = 0.25 (maximum possible proportion (0.5) * 1—maximum possible proportion (0.5) produces maximum possible sample size).
- d acceptable margin of error for proportion being estimated = 0.05 (error the researcher is willing to accept).

The sample size was thus calculated and was rounded to 250 respondents from the Information Technology sector. They were similar demographically to the respondents who participated in the pilot study.

4.7 Statistical Tools Used

After collecting the data, it needs to be analyzed in order to interpret the results and verify the correctness of the hypotheses. The quantitative analysis techniques provide helpful tools to analyze the data. Statistical package for Social Sciences (SPSS), now called Professional Analytic Software (PASW); Version 17.0 was used for statistical analysis of the collected and tabulated data. The following statistical techniques have been used for analysis across all the three research instruments. A brief description of these techniques is as follows:

4.7.1 Measure of Central Tendency (Mean)

This measure is mainly used for summarizing the essential features of a series and for enabling data to be compared. It is amiable to arithmetical treatment and is utilized as a part of further statistical calculations.

4.7.2 *Measures of Dispersion (Standard Deviation)*

An average can represent a series only to some extent and cannot reveal the entire story of any phenomenon under study. It fails to give any idea about the scatter of the values of items of a variable in the series around the true value of average. In order to measure this scatter, statistical dispersion is calculated.

4.7.3 *Percentage Analysis*

This technique measures the percentages of the demographic profile of the respondents which as discussed in the measurement scales, collectively comprises of five items, namely Gender, Age, Marital Status, Job Profile, Length of Stay overseas and Time Elapsed since Return from Overseas. The analysis threw light on the variability as far as the percentages of the sub-sections of the categories mentioned, is concerned.

4.7.4 *Correlation*

It is the most widely used statistic method for summarizing the strength of association between two metric (interval or ratio scaled) variables, say X and Y. It is an index used to determine whether a linear, or straight-line relationship exists between X and Y. It indicates the degree to which the variation in one variable, X, is related to the variation in another variable, Y. The correlation co-efficient '*r*' is also known as Pearson Correlation Co-efficient as it was originally proposed by Karl Pearson. It is also referred to as simple correlation, bi-variate correlation or merely the correlation co-efficient. Often Correlation analysis is used in conjunction with Regression analysis to measure how ell the regression line explains the variation of the dependent variable, Y. Correlation can also be used by itself, however to measure the degree of association between the two variables.

The correlations for mean scale scores indicated that dependent variable Anticipatory Adjustment and the constant variable Reverse Culture Shock are moderately correlated as the correlation coefficients were less than 0.5 ($p < 0.5$). The correlation matrix between various factors measured in the present study reveals that maximum correlation exists between Competency Transfer and Effective Repatriation (0.760) and correlation between Anticipatory Adjustment and Reverse Culture Shock is minimum (0.211) and negative. None of the correlations was higher than 0.8 which indicated that each variable can be distinct enough to measure different variables under the same construct (Sekaran, 2006).

The result indicated that dependent variables are significantly correlated with all independent variables. Moreover, all variables are also significantly correlated with each other with only low to moderate coefficient scores. Positive as well as negative correlation between the various constructs made a good research instrument for accessing the Antecedents of Repatriate Adjustment.

4.7.5 *Reliability Analysis of Measurement Scales*

Data Analysis is the most important task during the research process after data collection. The collected data should be scientifically and systematically processed and analyzed to achieve the stipulated objectives of the research, wherein the data processing involves editing, coding, classification and tabulation of gathered data. The calculations and computations of data are based on certain parameters and finding out the patterns of relationships that exist among various data groups. Thereafter, the testing of hypotheses with the help of statistical tools is done and subsequently the results are achieved.

Reliability is defined as the extent to which a Questionnaire, test, observation or any measurement procedure, produces the same results on repeated trials. The stability or consistency of scores after some time or crosswise over raters is ensured. Reliability is in this way the degree to which a trial, test or any measuring method yields the same results on rehashed trials.

The reliability of constructs was tested by using Cronbachs' coefficient alpha, which ranges from 0.791 to 0.855, much larger than the standard of 0.7 (Churchill, 1979). Since all measures were based upon prior relevant literature and were often used or adapted in research, evidence of content validity was provided. A pilot study for pre testing the Questionnaire was conducted and it validated the Cronbach alpha quotient reliability criteria (0.791–0.855).

The high level of Cronbachs Alpha for all the items of the four variables shows that it could be assumed that all items measure the respective construct. Since a Cronbachs Alpha of 0.7 or above is the ideal value that is considered reliable (DeVellis et al., 2003), all items in this study were very reliable because they have a good internal consistency, as shown in Table 4.1.

Table 4.1 Cronbachs alpha

Constructs	Cronbachs' alpha	Number of items
Competency transfer	0.855	12
Effective repatriation	0.839	13
Anticipatory Adjustment	0.798	12
Reverse culture shock	0.791	15

4.7.6 *Regression*

Regression analysis is a powerful and flexible statistical procedure for analyzing associative relationships between a metric dependent variable and one or more independent variables. It can be used in the following ways; To determine whether the independent variables explain a significant variation in the dependent variable; To determine how much how much of the variation in the dependent variable can be explained by the independent variables; To determine the structure or form of the relationship; To predict the values of the dependent variable; To analyze the control for other independent variables when evaluating the contributions of a specific variable or a set of variables.

Although the independent variables may explain the variation in the dependent variable, this does not necessarily imply causation. The use of the terms Dependent or criterion variables and Independent or predictor variables, in Regression Analysis arises from the mathematical relationship between the variables. These terms do not imply that the criterion variable is dependent on the independent variables in a causal sense. Regression analysis is concerned with the nature and degree of association between variables and does not imply or assume any causality. The mathematical equation is derived in the form of a straight line by using the least square procedure. When the regression is run on standardized data, the intercept assumes value of zero, and the regression co-efficients are called beta weights. The strength of association is measured by the Co-efficient of determination R^2

Further it is the operation of learning a function that predicts the value of a real valued dependent variable based on the values of other independent variables. It is a technique used for the modeling and analysis of numerical data consisting of values of a dependent variable and of one or more independent variables. The dependent variable in the regression equation is modeled as a function of the independent variables, corresponding parameters (constants) and an error term. The error term is treated as a random variable and represents unexplained variation in the dependent variable. The standard error of estimate is used to access the accuracy of prediction and maybe interpreted as a kind of average error made in predicting the dependent variable from the regression equation. Parameters are estimated to give a 'best-fit' of the data. Most commonly, the best fit is evaluated by using the least squares method, but other criteria have also been used.

For the present study, this tool has been used for studying the relationships among the various variables involved in the research process and hence the dependent and independent variables are assigned accordingly to the relationships to be analyzed.

4.7.7 *Factor Analysis and Validity*

Factor analysis is frequently used to develop a Questionnaire as it ensures that the questions asked relate to the construct which the researcher intends to measure. Factor analysis a method of data reduction, also called exploratory factor analysis (EFA), is a class of procedures used for reducing and summarizing data. Each variable is communicated as a linear blend of the fundamental components. Similarly, the factors themselves can be communicated as a linear mix of the observed variables. The factors are extracted in such a way that the first factor accounts for the highest variants in the data, the second the next highest, and so on. In formulating the factor analysis problem, the inclusion of variables in the analysis should be specifically based on previous research, theory and the judgment of the researcher. These variables then need to be measured on an interval or ratio scale, for arriving at results. Factor analysis is based on a matrix of correlation between the variables. In the process, the measures that are highly correlated, either positively or negatively, more likely get influenced by the same factors, while the relatively uncorrelated measures, more likely get influenced by different factors (DeCoster, 1998).

The appropriateness of the correlation matrix for factor analysis can be statistically tested. The two basic approaches to factor analysis are principal components analysis and common factor analysis. The present study used the principal component analysis method. Here the total variance in the data is considered. The diagonal of the correlation matrix comprises of unities, wherein the full variance is brought into the factor matrix. Principal component analysis is suggested when determination of the minimum number of factors that will account for maximum variance in the data for use in subsequent multivariate analysis, is the primary concern. The factors are called principal components. In principal component analysis the initial or unrotated factor matrix suggests the relationship between the factors and individual variables; it seldom results in factors that are interpretable, due to the fact that the factors are correlated with many variables. Therefore rotation is used to transform the factor matrix into a simpler one that is easier to interpret.

The most commonly used method of rotation is the Varimax procedure, which results in orthogonal factors. The rotated factor matrix forms the basis for interpreting the factors. Factor scores can be computed for each respondent. Alternatively, surrogate variables may be selected by examining the factor matrix and selecting for each factor, a variable with the highest or near highest loading. The differences between the observed correlations and the reproduced correlations, as estimated from the factor matrix, can be examined to determine model fit. The principal components method of extraction was used for data reduction, wherein the components with Eigen values more than 1 were extracted. The extracted components represented the variables well, because all the communalities were high.

In the present study, data was subjected to factor analysis for dimension reduction and for finding the relevant factor groups. In order to test the discriminant validity or the degree to which a study accurately reflects or assesses the specific concept which the researcher is attempting to measure or the extent to which the measures of conceptually distinct constructs differ (Sin et al., 2005), we conducted a Simple Factor Analysis with varimax rotation, using the Principal Component Method on fifty two initial items of the measures of Competency Transfer CT, Effective Repatriation ER, Anticipatory Adjustment AA and Reverse Culture Shock RCS, to determine the underlying dimensions. Purification of constructs determining the adjustment of

Table 4.2 KMO and bartletts test

Kaiser–Meyer–Olkin measure of sampling adequacy		0.768
Bartlett's test of sphercity	Approx. chi-square	4628.284
	df	820
	Sig	0.000

Repatriates was carried out using SPSS (Version 17.00). The process brought the constructs to the level of forty one statements of four factor groups, out of fifty two statements, originally kept in the domain of access.

Results indicated that four factors emerged, which could explain 50.403% of the total variance. Factor 4 carried 50.40% of data, indicating that most of the data fits into that factor. The items with highest loadings were used in assigning the labeling of the factor. As per the results of Factor Analysis, the first four factors were chosen because they explained a high proportion of original variance and had Eigen value higher than one. As depicted in Table 4.2, Bartlett's Test turned out to be significant; the KMO being 0.768 and Factor loadings ranging between 0.420 and 0.793, which indicated a high validity of the measurements and reflected that our respondents could discriminate among the measures of Competency Transfer, Effective Repatriation, Anticipatory Adjustment and Reverse Culture Shock, hence offering evidence of discriminant validity (Podsakoff & Organ, 1986).

Exploratory Factor Analysis was used through SPSS (17 version) with Principal Component Analysis along with Varimax rotation for summarization of the total data falling under the construct of "Reverse Culture Shock", measured through the research instrument (Questionnaire), to arrive at meaningful factor groups. Such grouping of the statements of variables pertaining to the measurement of Reverse Culture Shock, in distinct areas of importance, yielded two factor groups; "Alienation and Segregation" comprising of four statements and "Loneliness and Anxiety" comprising of five statements, covering different dimensions of Reverse Culture Shock experienced by Repatriates. The implications for these factors are in relevance to the importance of these factors in impacting Anticipatory Repatriate Adjustment. Similarly factor analysis of the total data falling under the construct of "Effective Repatriation", measured through the research instrument (Questionnaire), converged at two factor groups; "Career aspects" (five statements) and "Acumen, Communication, Changes and Repatriation Stress" (four statements), covering different dimensions of Effective Repatriation(ER) of Repatriates.

The implications for these factors have been discussed in relevance to the importance of these factors in impacting the dependent variables "Anticipatory Repatriate Adjustment" and "Competency Transfer". Subsequently, correlations and regression between the factor groups and dependent variables were explored.

Validity is thus assessed by determining whether each indicator's estimated pattern coefficient on its posited underlying construct factor is significant (Anderson & Gerbing, 1988). To determine the constructs' validity, the path coefficients (i.e., factor loadings) from the latent constructs to the corresponding indicators for all the constructs were analyzed (Sujan et al., 1994). The results indicated that all the factor loadings were positive and all items were significantly loaded to the intended factors. Moreover, the results indicated that if the data is subjected to factor analysis, there will be no or little multiple loading in the rotated factor loading table. Hence the data was fit for factor analysis.

4.7.8 Anova

The analysis of variance or ANOVA is used for examining the differences in the mean values of the dependent variable associated with the effect of the controlled independent variables, after taking into account the influence of the uncontrolled independent variables. Essentially, ANOVA in SPSS is used as the test of means for two or more populations.

ANOVA must have a dependent variable which should be metric (measured using an interval or ratio scale). ANOVA must also have one or more independent variables, which should be categorical in nature. In ANOVA, categorical independent variables are called factors. A particular combination of factor levels or categories is called a treatment. In ANOVA, there is one way ANOVA which involves only one categorical variable, or single factor. In one way ANOVA in SPSS, a treatment is the same as the factor level. If two or more factors are involved in ANOVA, then it is termed as '*n*' way ANOVA.

In the present study, data was subjected to ANOVA test in order to analyze the variations in one variable with respect to the other variables involved herein.

References

Anderson, J. C., & Gerbing, D. W. (1988). Structural equation modeling in practice: A review and recommended two-step approach. *Psychological Bulletin, 103*(3), 411–423.

Andersson, J. (2006). *Effective repatriation: A case study of volvo construction equipment in Eskilstuna*. diva-portal.org.

Ary, D., Jacobs, L. C., & Razavieh, A. (1996). *Introduction to research in education* (5th ed.). Harcourt Brace College Publishers.

Black, J. S., & Gregersen, H. B. (1991). When Yankee comes home: Factors related to expatriate & spouse repatriation adjustment. *Journal of International Business Studies, 22*(4), 671–694.

Black, J. S., Gregersen, H. B., & Mendenhall, M. E. (1992). Toward a theoretical framework of repatriation adjustment. *Journal of International Business Studies, 23*(4), 737–760.

Black, J. S., Gregersen, H. B., Mendenhall, M. E., & Stroh, L. (1999a). *Globalizing people through international assignments*. Addison-Wesley Longman.

Black, J. S., Gregersen, H. B., & Morrison, A. (1999b). *Global explorers: The next generation of leaders*. Routledge.

Brislin, R. W. (1981). Issues of effective repatriation: A model and managerial implications. *Cross-Cultural Encounters, 45*(3), 56–69.

Churchill, G. A. (1979). A paradigm for developing better measures of marketing constructs. *Journal of Marketing Research, 16*(1), 64–73.

Cochran, W. G. (1977). *Sampling techniques* (3rd ed.). Wiley.
Cooper, D. R., Schindler, P. S., & Sun, J. (2006). *Business research methods*. sutlib2.sut.ac.th.
DeCoster, J. (1998). *Overview of factor analysis*. Department of Psychology, University of Alabama. stat-help.com.
DeVellis, R. F., Lewis, M. A., & Katherine, R. S. (2003). Interpersonal emotional processes in adjustment to chronic illness. *Social Psychological Foundations of Health and Illness*, books.google.com.
Furuya, N., Stevens, M. J., Oddou, G., Bird, A., & Mendenhall, M. E. (2007). The effects of HR policies and repatriate self-adjustment on global competency transfer. *Asia Pacific Journal of Human Resources, 45*(1), 6–23.
Furuya, N., Stevens, M., Bird, A., Oddou, G., & Mendenhall, M. (2008). Managing the learning & transfer of global management competence: Antecedents & outcomes of Japanese repatriation effectiveness. *Journal of International Business Studies, 40*, 200–215.
Gaw, K. F. (2000). Reverse culture shock in students returning from overseas. *International Journal of Intercultural Relations, 24*(1), 83–104.
Hermansson, F., & Andersson, R. (2008). *Assimilation of repatriate knowledge: A study of knowledge management in 39 Large Cap companies*. Disciplinary domain of Humanities and Social Sciences, Faculty of Social Sciences, Department of Business Studies, Uppsala University.
Hermansson, F., & Kilnes, U. (2008). *Knowledge transfer from expatriates: A study of MNCs' exploitation of expatriates' knowledge*. Uppsala University, Faculty of Social Sciences, Department of Business Studies.
Jassawalla, A., Traci, C., & Lindsay, S. (2004a). Practices imperative to management of human resources. *S.A.M. Advanced Management Journal, 69*(2), 38–46.
Jassawalla, A., Truglia, C., & Garvey, J. (2004b). Cross-cultural conflict and expatriate manager adjustment: An exploratory study. *Management Decision, 42*(7), 837–849.
Kothari, C. R. (2004). *Research methodology: Methods and techniques*. books.google.com.
Krejcie, R. V., & Morgan, D. W. (1970). Determining sample size for research activities. *Educational and Psychological Measurement, 30*, 607–610.
Leedy, P. D., & Ormrod, J. E. (2005). *Practical research: Planning and design*.
Löyvä, N. (2014). The repatriation process of finnish international assignees. *General Management*. University of Sweden.
Martin, D. C., & Anthony, J. J. (2006). The repatriation and retention of employees: Factors leading to successful programs. *International Journal of Management, 23*(3), 620–631.
Masako, L. (2008). The nature of reentry shock of women: A survey among the wives of Japanese corporate sojourners. *Okinawa Junior College Review, Okinawa University, 13*(1), 57–121.
Murray, O. T. (2007). *The effects of repatriation on managers returning from foreign assignments*. Dublin Business School.
Niesen, C. C. (2010). *Navigating reentry shock: The use of communication as a facilitative tool*. University of New Mexico, Department of Communication & Journalism.
Paganus, S. (2005). *Finish business repatriates coping strategies*. Swedish School of Economics and Business Administration, Department of Management and Organisation, Economics and Society—154, Helsingfors.
Podsakoff, P. M., & Organ, D. W. (1986). Self-reports in organizational research: Problems and prospects. *Journal of Management, 12*(4), 531–544.
Santosh, B. R., & Muthiah, K. (2011). Repatriation program as a process for retaining repatriates—Issues surrounding international human resource. *International Journal of Research in Commerce, IT & Management, 1*(3).
Santosh, B. R., & Muthiah, K. (2012). Knowledge transfer from repatriated employees: The Indian Experience. *IUP Journal of Knowledge Management, 10*(1), 7–25.
Saunders, M., Lewis, P., & Thornhill, A. (2011). *Research methods for business students*. books.google.com.
Sekaran, U. (2006). *Research methods for business: A skill building approach*, 4Th Ed. books.google.com.

Siddiqua, A. (2010). Reverse culture shock among returnee high school students. *Malaysian Journal of Psychiatry, 19*(1).

Sin, L. Y. M., Tse, A. C. B., & Yim, F. H. K. (2005). CRM: Conceptualization and Scale development. *European Journal of Marketing, 39*(11), 1264–1290.

Sujan, H., Weitz, B. A., & Kumar, N. (1994). Learning orientation, working smart, and effective selling. *Journal of Marketing, 58*(3), 39–52.

Torbiorn, I. (1982). *Living abroad: Personal adjustment and personnel policy in the overseas setting.* John Wiley & Sons.

Chapter 5
Data Analysis and Interpretation

5.1 Introduction

Data gathered by the researcher must be scientifically and empirically handled or analyzed in order to accomplish the objectives of the research. Data analysis includes altering, coding, classifying and organizing of the gathered information with the goal that it is credible, ready and competent for analysis in all regards. Data analysis is referred to the computations and calculations of information in light of certain parameters, basis the pattern of relationship that exists among the data groups, subsequently tested with hypotheses of the study through support of statistical tools. The present chapter provides empirical findings and analysis based on the data collected through the research questionnaire, in order to test the hypotheses and to provide answers to the research questions.

The SPSS software (now called PASW) version 17 was used to analyze and interpret the data and transform it into information. This chapter starts with discussing the descriptive analysis of respondents, reliability and validity test, factor analysis, correlation analysis, regression, t-test, ANOVA and discussion of the findings.

5.2 Demographic Descriptive Analysis

Descriptive analysis is used to gain a better understanding of the collected data (Runeson & Host, 2009). This section presents the statistical sample distribution with regard to the variables such as, gender, age, marital status, job profile, length of stay overseas and time elapsed since return from overseas assignment.

M. B. Nag and F. Ahmad Malik, *Repatriation Management and Competency Transfer in a Culturally Dynamic World*,
https://doi.org/10.1007/978-981-19-7350-5_5

Table 5.1 Gender profile of the respondents

Gender	Frequency	Percent (%)
Male	205	82
Female	45	18
Total	250	100

Table 5.2 Age-wise representation of respondents

Age	Frequency	Percent (%)
20–29	69	27.6
30–39	105	42
40–49	65	26
50–59	11	4.4
Total	250	100

5.2.1 Gender

In terms of gender distribution, there is an unequal representation of both the genders. As demonstrated in Table 5.1, the respondents include 82% males and 18% females, out of the total sample of 250 respondents.

5.2.2 Age

The respondents are distributed in four age groups as illustrated in Table 5.2. The age structure of the respondents reveals that the majority of them were in the age group 30–39 years representing 42% of the total sample.

5.2.3 Marital Status

Table 5.3 exhibits the marital status of the respondents, primarily comprising 75.2% married, 21.6% single and 3.2% belonging to unspecific other category.

Table 5.3 Marital status of respondents

Marital status	Frequency	Percent (%)
Single	54	21.6
Married	188	75.2
Others	08	3.2
Total	250	100

Table 5.4 Length of stay overseas

Length of stay	Frequency	Percent (%)
< 6 months	64	25.6
7–12 months	85	34
13–24 months	58	23.2
> 25 months	43	17.2
Total	250	100

Table 5.5 Job profile of respondents after repatriation

Job profile	Frequency	Percent (%)
Top level	16	6.4
Middle level	174	69.6
Lower level	60	24
Total	250	100

5.2.4 Length of Stay Overseas

Table 5.4 illustrates the duration of stay of respondents in overseas locations before repatriating back to home country. Results reveal that 34% stayed for 7–12 months, followed by 25.6% who stayed overseas for less than six months, 23.2% who stayed for 13–24 months and 17.2% who stayed for more than 25 months.

5.2.5 Job Profile of Respondents

Table 5.5 shows the Job profiles of the respondents, once they repatriated back to the home country. Majority of the respondents belong to the middle level of hierarchy comprising 69.6% followed by 24% in the lower level of hierarchy and only 6.4% of the total number of respondents belong to the top level of hierarchy.

5.2.6 Time Elapsed Since Return

As illustrated in Table 5.6, the time elapsed since return from overseas assignment was divided into four broad categories and the highest percentage of 29.2% was recorded in the time bracket of 7–12 months, followed by 27.2% (> 19 months), 24.8%(13–18 months) and 18.8% (< 6 months).

Table 5.6 Time elapsed since return

Time elapsed since return	Frequency	Percent (%)
< 6	47	18.8
7–12 months	73	29.2
13–18 months	62	24.8
> 19 months	68	27.2
Total	250	100

5.3 Reliability and Validity

Reliability and validity were considered carefully in this study. It refers to the degree to which measures are free from error and therefore consistent results from the study are yielded (Zikmund & Babin, 1997; Malhotra et al., 1996). Validity refers to the accuracy of measurement, whether the conceptual and operational definitions are truly a reflection of the underlying concept to be measured (Burns & Bush, 1995; Newman & Newman, 1994). Cronbach's Alpha is used to measure the internal consistency and reliability and is considered to be the most common measure of reliability (Roberts et al., 2006). The present study consists of four variables measured with 52 items. Cronbach's Alpha (α) was used to check the reliability of these variables. Most research methods consider a measure with Cronbach's Alpha equal to or greater than 0.7 as reliable (Gliem & Gliem, 2003). Reliability refers to the correlation on a variable; high correlation means that the results are consistent and therefore are reliable.

By using Cronbach's Alpha coefficients, the internal reliability for all items of each construct was first calculated. As shown in Table 5.7, reliabilities of each construct ranged from 0.791 to 0.855. The high level of Cronbach's Alpha for all the items of the four constructs, show that it could be assumed, that all items measure the respective construct. Since a Cronbach's Alpha of 0.7 or above is the ideal value that is considered reliable (DeVellis, 2003), all constructs in this study are very reliable as they have good internal consistency.

Table 5.7 Cronbachs alpha

Constructs	Cronbachs alpha	Number of items
Competency transfer	0.855	12
Effective repatriation	0.839	13
Anticipatory adjustment	0.798	12
Reverse culture shock	0.791	15

5.4 Factor Analysis

Factor analysis is frequently used to develop a Questionnaire as it ensures that the questions asked relate to the construct which the researcher intends to measure. Factor analysis a method of data reduction, also called exploratory factor analysis (EFA), is a class of procedures used for reducing and summarizing data. Every variable is communicated as a linear blend of the fundamental components. Similarly, the components themselves can be communicated as combinations of observed variables. The factors are extracted in such a way that the first factor accounts for the highest variants in the data, the second the next highest, and so on. In formulating the factor analysis problem, the variables to be included in the analysis should be specified based on past research, theory and the judgment of the researcher. These variables should be measured on an interval or ratio scale. Factor analysis is based on a matrix of correlation between the variables. Measures that are highly correlated, either positively or negatively are likely to get influenced by the same factors, while those that are relatively uncorrelated are likely influenced by different factors (DeCoster, 1998).

The appropriateness of the correlation matrix for factor analysis can be statistically tested. The two basic approaches to factor analysis are principal components analysis and common factor analysis. The present study used the principal component analysis method. Here the total variance in the data is considered. The diagonal of the correlation matrix consists of unities, and full variance is brought into the factor matrix.

Principal component analysis is recommended when the primary concern is to determine the minimum number of factors that will account for maximum variance in the data for use in subsequent multivariate analysis. The factors are called principal components. In principal component analysis the initial or un-rotated factor matrix indicates the relationship between the factors and individual variables; it seldom results in factors that can be interpreted, because the factors are correlated with many variables. Therefore rotation is used to transform the factor matrix into a simpler one that is easier to interpret. The most commonly used method of rotation is the Varimax procedure, which results in orthogonal factors. The rotated factor matrix forms the basis for interpreting the factors. Factor scores can be computed for each respondent. Alternatively, surrogate variables may be selected by examining the factor matrix and selecting for each factor, a variable with the highest or near highest loading. The differences between the observed correlations and the reproduced correlations, as estimated from the factor matrix, can be examined to determine model fit. The principal components method of extraction was used for data reduction. Components with Eigen values greater than 1 were extracted. As the communalities were all high, the extracted components represented the variables well.

Validity is assessed by "determining whether each indicator's estimated pattern coefficient on its posited underlying construct factor was significant" (Anderson & Gerbing, 1988). Data is subjected to factor analysis for dimension reduction and for finding relevant factors. To test validity, we conducted a simple factor analysis using the principal component method on measures of Competency Transfer CT, Effective Repatriation ER, Anticipatory Adjustment AA and Reverse Culture Shock, RCS (Table 5.8). Purification of constructs determining the adjustment of Repatriates was carried out using SPSS (Version 17.00) and the emergent results are shown in Table 5.8. Constructs under consideration are Competency Transfer, Anticipatory Adjustment, and Effective Repatriation and Reverse Culture Shock. In order to assess the perceptions of respondents, a Factor Analysis with varimax rotation on the 52 initial items categorized the items into the four groups of "Competency Transfer", "Anticipatory Adjustment", "Effective Repatriation" and "Reverse Culture Shock", was performed. The perceived importance of the 52 statements was factor analyzed; using principal component analysis with Varimax rotation to determine the underlying dimensions (Table 5.8). The process of Principal Component Analysis (PSA) with Varimax Rotation brought the constructs to the level of 34 statements out of 52 statements originally kept in the domain of access.

Therefore, Factor Analysis reduced the initial 52 items into 34 items of 4 factor groups. The items with highest loadings were used in assigning the labeling of the factor. As per the results of Factor Analysis, the first four factors were chosen because they explained a high proportion of original variance and had Eigen value higher than one. Factor loadings range between 0.435 and 0.781 (Table 5.8) and significant at 0.05 levels, indicating high validity of the measurements and reflecting that our respondents could discriminate among all the measures, offering evidence of validity (Podsakoff & Organ, 1986). To determine the constructs' validity, the path coefficients (i.e., factor loadings) from the latent constructs to the corresponding indicators for all the constructs were analyzed (Sujan et al., 1994). The results in Table 5.8 show the factor loadings for each item in relation to the various factors and indicate that all the factor loadings are positive and all items are significantly loaded to the intended factors. Moreover, the results indicated that when the data will be subjected to factor analysis, there will be no or little multiple loading in the rotated factor loading table. Hence the data is fit for factor analysis.

Results of the factor analysis indicated that the four factors emerged, which could explain 50.403% of the total variance (Table 5.9). Table 5.9 shows how much of the total data fits into the four factors and this is calculated using variance. The total cumulative variance percentage accumulated in the resultant four constructs is 50.40% and factor 4 carries 50.40% of data, indicating that most of the data fits into that factor. The other three factors carry lesser percentage each and show relatively low fit of data in the factors. The communalities and percentage of variance explained by each factor is displayed in the Table 5.9. The four constructs in order of sequence, exhibited the following values.

Extraction Method: Principal Component Analysis

The suitability of raw information for factor examination got from respondents is inspected through KMO value, Bartlett's Test of Sphercity and p-value, showing adequate basic variance and relationship matrix (Dess et al., 1997). The KMO value (0.768) and Bartlett test of Sphercity (4628.284) indicates acceptable and significant values (Table 5.10).

Table 5.8 Rotated component matrix

	Component			
	1	2	3	4
CT02	0.736			
CT04	0.844			
CT05	0.781			
CT06	0.805			
CT07	0.761			
CT08	0.760			
CT09	0.654			
CT11	0.699			
ER1		0.436		
ER2		0.591		
ER3		0.687		
ER4		0.512		
ER5		0.479		
ER7		0.441		
ER8		0.539		
ER9		0.495		
ER11		0.514		
AA2			0.441	
AA4			0.494	
AA5			0.543	
AA6			0.487	
AA9			0.561	
A10			0.654	
AA11			0.641	
AA12			0.586	
RCS1				0.472
RCS2				0.576
RCS3				0.435
RCS4				0.486
RCS6				0.569
RCS7				0.537
RCS9				0.535
RCS8				0.511
RCS11				0.532

Table 5.9 Total variance explained

Component	Initial Eigen values			Extraction sums of squared loadings			Rotation sums of squared loadings		
	Total	% of variance	Cumulative %	Total	% of variance	Cumulative %	Total	% of variance	Cumulative %
1	10.696	26.088	26.088	10.696	26.088	26.088	9.140	22.294	22.294
2	4.104	10.009	36.097	4.104	10.009	36.097	4.825	11.768	34.061
3	3.633	8.861	44.958	3.633	8.861	44.958	4.107	10.017	44.078
4	2.233	5.445	50.403	2.233	5.445	50.403	2.593	6.325	50.403
5	1.908	4.654	55.057						
6	1.806	4.404	59.461						
7	1.594	3.889	63.350						
8	1.328	3.238	66.588						
9	1.270	3.097	69.686						
10	1.094	2.668	72.353						
11	1.011	2.466	74.820						
12	0.930	2.268	77.088						
13	0.839	2.046	79.134						
14	0.745	1.817	80.950						
15	0.698	1.702	82.652						
16	0.635	1.549	84.201						
17	0.587	1.431	85.633						

(continued)

Table 5.9 (continued)

Component	Initial Eigen values			Extraction sums of squared loadings			Rotation sums of squared loadings		
	Total	% of variance	Cumulative %	Total	% of variance	Cumulative %	Total	% of variance	Cumulative %
18	0.525	1.282	86.915						
19	0.502	1.225	88.139						
20	0.457	1.114	89.253						
21	0.441	1.074	90.328						
22	0.413	1.007	91.335						
23	0.373	0.909	92.244						
24	0.361	0.880	93.125						
25	0.326	0.796	93.921						
26	0.295	0.720	94.641						
27	0.257	0.627	95.267						
28	0.238	0.581	95.848						
29	0.209	0.509	96.358						
30	0.194	0.474	96.832						
31	0.182	0.445	97.277						
32	0.164	0.401	97.678						
33	0.156	0.379	98.057						
34	0.138	0.338	98.395						
35	0.129	0.315	98.710						
36	0.110	0.268	98.978						
37	0.107	0.261	99.239						

(continued)

Table 5.9 (continued)

Component	Initial Eigen values			Extraction sums of squared loadings			Rotation sums of squared loadings		
	Total	% of variance	Cumulative %	Total	% of variance	Cumulative %	Total	% of variance	Cumulative %
38	0.102	0.249	99.488						
39	0.089	0.217	99.705						
40	0.071	0.173	99.879						
41	0.050	0.121	100.000						

Table 5.10 KMO and Bartlett's test

Kaiser–Meyer–Olkin	Measure of sampling adequacy	0.768
Bartlett's test of sphercity	Approx. Chi-Square	4628.284
	df	820
	Sig	0.000

KMO is calculated using correlation and partial correlation to test whether the variables in the sample are adequate to correlate or not (Table 5.10). The KMO test varies between 0 and 1 and the value of 0 shows that the sum of partial correlations is large relative to the sum of correlations, meaning that the factor analysis is likely to be irrelevant while a value close to 1, shows that the patterns of correlations are relatively compact and factor analysis yields distinct and reliable factors (Field, 2005). This test proves that the data does not have multi-co linearity and hence can be subjected to factor analysis.

As a rule of thumb, the value of KMO should be more than 0.5 to proceed for factor analysis. For our case, the KMO value is 0.768 (Table 5.10), which indicates that factor analysis is relevant for our study and there is no multicollinearity in the data. Bartlett's Test of Sphericity helps to know whether there are relationships between the variables. A p-value less than 0.5 indicates that it makes sense to continue with factor analysis and as shown in Table 5.9, its value is 0.000 and hence is significant.

Therefore the Kaiser–Meyer–Olkin test of sphericity and the Bartlett's test of adequacy provided support for the factor analysis. Furthermore, the prerequisites followed in this study were the accompanying: Eigenvalues > 1, cut-off points > 0.40, cross-loadings > 0.10, and Cronbach's alpha > 0.70. Thus, pursuant to the results of factor analysis, this study successfully identified the underlying dimensions of Repatriate adjustment as perceived by the Repatriates who had moved back to India.

Therefore, factor loading in the final factorial design are steady with traditionalist criteria, arriving at four-factor arrangement, utilizing Kaiser Criteria (i.e. Eigen value $\geq$ 1) with 50.40% of the total variance explained.

- Factor 1: Competency Transfer (Value = 26.088, Cronbachs alpha = 0.855)
- Factor 2: Effective Repatriation (Value = 36.097, Cronbachs alpha = 0.839)
- Factor 3: Anticipatory Adjustment (Value = 44.958, Cronbachs alpha = 0.798)
- Factor 4: Reverse Culture Shock (Value = 50.403, Cronbachs alpha = 0.791)

5.4.1 *Items Under Constructs After Applying Factor Analysis*

Factor 1, labeled as "Competency Transfer" now consisted of eight statements out of initial twelve statements after applying the Rotated component matrix tool wherein four items with weaker factor loadings (CT1, CT3, CT10 and CT12) got eliminated. The filtered statements comprised of Skill and acumen utilization (CT2), Knowledge sharing with co-workers (CT5), Skill transfer (CT6), Hold seminars (CT7),

Organizational benefits (CT8), Career rewards (CT9), Decision making in International projects (CT11) and Job position and Skill utilization (CT4). The item with the highest factor loading is CT4 (0.844) i.e. Job position and skill utilization (Table 5.8).

Factor 2, labeled as "Effective Repatriation" now includes nine items i.e. Repatriation training and benefits (ER1), Job position and hierarchy (ER2), Job description (ER3), Career progression ahead of colleagues (ER4), Role of communication in uncertainty elimination (ER5), Social status post repatriation (ER7), Acquired acumen and career development (ER8), No repatriation stress (ER9) and Effects of changes in home country (ER11). The item with the highest factor loading is ER3 (Job description) i.e. 0.687 (Table 5.8). ER6, ER10, ER12 and ER13 got eliminated due to either cross loadings and or zero factor loadings.

Factor 3, is labeled as "Anticipatory Adjustment" and finally consists of eight statements after the initial twelve statements. These are Repatriation expenses (AA2), Smooth family adjustment (AA4), Role clarity and hierarchy of Job (AA5), Lesser role conflict (AA6), Improved social status (AA9), Favorable housing conditions (AA10), No effect of changes back home (AA11), Smooth personal repatriation adjustment (AA12). AA1, AA3, AA7 and AA8 got eliminated. The item with the highest factor loading is favorable housing conditions AA10 i.e. 0.654 (Table 5.8).

Factor 4, labeled a "Reverse Culture Shock" got filtered to nine items from initial fifteen items. The statements now comprised of Pre-Repatriation anxiety (RCS1), Clash of expectations and reality (RCS2), Problems in communication post Repatriation (RCS3), Sense of frustration, Pre and Post Repatriation (RCS4), Depression post Repatriation (RCS6), Lonely and isolated post Repatriation (RCS7), Alienation post Repatriation (RCS8), Perceived Changes in values and attitudes of home country nationals post Repatriation (RCS9), Missing foreign culture (RCS11). The item with the highest factor loading is RCS 2 (0.576) i.e. Clash of expectation and reality (Table 5.8).

The four empirically derived factors coincide with the conceptually derived perceptions of respondents towards Repatriate adjustment discussed in review of literature. Thus, we can conclude that the conceptually derived variables (items) have been empirically validated with the four factors in the Antecedents of Repatriate adjustment context. Contrasting these outcomes with other comparable global examination information lies past the extent and scope of this study, however such an errand could be viewed as a valid contribution to the academic discussion regarding the view of Repatriation adjustment.

In light of these results we may accept that there is a set of constructs as measured in the current study which favorably analyzes the context of Repatriation adjustment.

5.5 Descriptive Statistics for Various Measurement Items

The following section specifies the factor loadings, mean values and the respective standard deviations of the various measurement items of the present study.

5.5.1 *Respondents' Perception Regarding Competency Transfer Items*

The suitability of raw information for factor examination acquired from respondents was inspected through KMO value, Bartlett's Test of Sphercity and *p*-value, (Table 5.10) demonstrating adequate normal variance and relationship matrix (Dess et al., 1997). The process of Principal Component Analysis (PCA) with Varimax Rotation brought the construct to the level of eight statements out of twelve statements originally kept in the domain of access. The communality for all the statements was above 0.5, indicating moderate to high degree of linear association among the variables. The factor loading ranges from 0.654 to 0.844 (Table 5.12). Table 5.11 indicates that the total cumulative variance explained by the factor Competency Transfer of Repatriates is 57.332%. The mean values varied between 3.10 and 4.01 and the total mean value stands at 3.5414. The total computed standard deviation of the items stands at 0.828. On the whole, all selected items significantly contribute towards this construct, thereby suggesting that the subjects believe that the congruity between Competency Transfer and Repatriate Adjustment is important as the overall mean value of 3.54 on a 5-point scale clearly indicates its relevance. This implies that most respondents perceived these eight statements as being of moderate importance in general. The mean values of items of Competency Transfer are approximately equal. The standard deviation is 0.828 which means that the scores are spread away from the mean. Overall, the results indicated that the respondents viewed logical connection and fit between Competency Transfer and the Repatriate Adjustment. The factor loading of each item indicates that the measurement scales of the constructs were stable and consistent.

Factor loadings are the weights and correlations between each variable and the factor. These values show the weight and correlation each item has to a factor or component. The higher the load, the more important it is in defining the factor's dimensionality. A negative value indicates an inverse impact on the factor.

Table 5.11 Total variance explained (competency transfer)

Component	Initial Eigen values			Extraction sums of squared loadings		
	Total	% of variance	Cumulative %	Total	% of variance	Cumulative %
1	4.587	57.332	57.332	4.587	57.332	57.332
2	0.834	10.419	67.751			
3	0.668	8.345	76.096			
4	0.580	7.244	83.340			
5	0.440	5.500	88.840			
6	0.368	4.604	93.444			
7	0.279	3.488	96.932			
8	0.245	3.068	100.000			

Extraction method Principal component analysis

Table 5.12 Analysis of competency transfer items

Competency transfer	Factor load	Mean	Standard deviation
CT4-Job position and skill utilization	0.844	3.7063	1.04969
CT6-Skill transfer	0.805	3.3813	1.13753
CT5-Knowledge sharing with co-workers	0.781	3.5063	1.20271
CT7-Hold seminars	0.761	3.1000	1.19327
CT8-Organizational benefit	0.760	3.4625	1.03333
CT2-Skill and acumen utilization	0.736	4.0125	0.94495
CT11-Decision making in international Projects	0.699	3.7313	1.04443
CT9-Career rewards	0.654	3.4313	1.13615
Overall		3.5414	0.82815

Extraction method Principal component analysis

Factor loadings range between 0.654 and 0.844 (Table 5.12) and significant at 0.05 levels, indicating high validity of the measurements and reflecting that our respondents could discriminate among all the measures, offering evidence of validity (Podsakoff & Organ, 1986). The highest factor loading was 0.844 for the statement pertaining to Job position and skill utilization.

To determine the constructs' validity, the path coefficients (i.e., factor loadings) from the latent constructs to the corresponding indicators for all the constructs were analyzed (Sujan et al., 1994). The results in Table 5.12 show the factor loadings for each item in relation to the various factors and indicate that all the factor loadings are positive and all items are significantly loaded to the intended factors. Moreover, the results indicated that when the data will be subjected to factor analysis, there will be no or little multiple loading in the rotated factor loading table.

5.5.2 Respondents' Perception Regarding Effective Repatriation Items

The suitability of raw information for factor examination acquired from respondents was inspected through KMO value, Bartlett's Test of Sphercity and *p*-value, (Table 5.10) demonstrating adequate normal variance and relationship matrix (Dess et al., 1997). The process of Principal Component Analysis (PCA) with Varimax Rotation brought the construct to the level of ten statements out of thirteen statements originally kept in the domain of access. The communality for the nine statements was above 0.5, indicating moderate to high degree of linear association among the variables. The factor loading ranges from 0.436 to 0.687 (Table 5.14) and significant at 0.05 levels, indicating high validity of the measurements and reflecting that our respondents could discriminate among all the measures, offering evidence of validity (Podsakoff & Organ, 1986). Table 5.13 indicates that the total cumulative variance explained by the factor Effective Repatriation of Repatriates is 49.932%. The mean values varied between 3.0438 and 3.7688 and the total mean value stands at 3.4139. The total computed standard deviation of the items stands at 0.82815.

On the whole, all selected items significantly contribute towards this construct, thereby suggesting, the subjects believe that the congruity between Effective Repatriation and Repatriate Adjustment is important as the overall mean value of 3.4139 on a 5-point scale clearly indicates its relevance. This implies that most respondents perceived these ten statements as being of moderate importance in general. The mean values of items of Effective Repatriation are approximately equal. The standard deviation is 0.828, which means that the scores are spread away from the

Table 5.13 Total variance explained (effective repatriation)

Component	Initial Eigen values			Extraction sums of squared loadings		
	Total	% of variance	Cumulative %	Total	% of variance	Cumulative %
1	4.494	49.932	49.932	4.494	49.932	49.932
2	1.019	11.323	61.255			
3	0.863	9.594	70.848			
4	0.752	8.359	79.207			
5	0.567	6.298	85.505			
6	0.452	5.017	90.522			
7	0.406	4.511	95.032			
8	0.248	2.759	97.791			
9	0.199	2.209	100.000			

Extraction method Principal component analysis

Table 5.14 Analysis of effective repatriation items

Effective repatriation	Factor load	Mean	Standard deviation
ER3-Job description	0.687	3.4625	1.02108
ER2-Job position and hierarchy	0.591	3.4438	0.99525
ER8-Acquired acumen and career development	0.539	3.7688	1.00453
ER11-Effect of changes in home country	0.514	3.4063	1.04187
ER4-Career progression ahead of colleagues	0.512	3.4125	0.98023
ER9-No repatriation stress	0.495	3.1313	1.13892
ER5-Role of communication in uncertainty elimination	0.479	3.6875	0.93961
ER7-Social Status post repatriation	0.441	3.0438	1.12907
ER1-Repatriation training and benefit	0.436	3.3688	1.03171
Overall		3.4139	0.72835

Extraction method Principal component analysis

mean. Overall, the results indicated that the respondents viewed logical connection and fit between the Effective Repatriation and the Repatriate Adjustment. The factor loading of each item indicates that the measurement scales of the constructs were stable and consistent.

To determine the constructs' validity, the path coefficients (i.e., factor loadings) from the latent constructs to the corresponding indicators for all the constructs were analyzed (Sujan et al., 1994). The results in Table 5.12 show the factor loadings for each item in relation to the various factors and indicate that all the factor loadings are positive and all items are significantly loaded to the intended factors. Moreover, the results indicated that when the data will be subjected to factor analysis, there will be no or little multiple loading in the rotated factor loading table.

5.5.3 Respondents' Perception Regarding Anticipatory Adjustment Items

The suitability of raw data for factor analysis obtained from respondents is examined through KMO value, Bartlett's test of Sphercity and p-value, (Table 5.10) indicating sufficient common variance and correlation matrix (Dess et al., 1997). The process of Principal Component Analysis (PCA) with Varimax Rotation brought the construct to the level of eight statements out of twelve statements originally kept in the domain of access. The communality for the eight statements was above 0.5, indicating moderate to high degree of linear association among the variables. The factor loading ranges from 0.441 to 0.654 (Table 5.16), indicating that the factor loading values of the eight items of point towards relatively fair importance of items in defining the dimensions of Anticipatory Adjustment. Table 5.15 indicates that the total cumulative variance explained by the factor Anticipatory Adjustment of Repatriates is 45.078%. The mean values varied between 3.6500 and 4.1063 and the total mean value stands at 3.8000, which on a five point scale indicate the varying familiarity of the Repatriates to Anticipatory Adjustment. The total computed standard deviation of the items stands at 0.72835.

On the whole, all selected items significantly contribute towards this construct, thereby suggesting, the subjects believe that the congruity between Anticipatory Adjustment and Repatriate Adjustment is important as the overall mean value of 3.8000 on a 5-point scale clearly indicates its relevance. This implies that most respondents perceived these eight statements as being of importance in general. The mean of the responses to Anticipatory Adjustment factor item "Smooth personal Repatriate adjustment" is 4.10 and that of "Role clarity and hierarchy of job" is 3.65. The results point out that a majority of respondents "agree" that they are familiar with the specified items of Anticipatory Adjustment.

The mean values of items of Anticipatory Adjustment are approximately equal. The standard deviation is 0.728, which means that the scores are spread away from the mean. Overall, the results indicated that the respondents viewed logical connection

Table 5.15 Total variance explained (anticipatory adjustment)

Component	Initial Eigen values			Extraction sums of squared loadings		
	Total	% of variance	Cumulative %	Total	% of variance	Cumulative %
1	3.606	45.078	45.078	3.606	45.078	45.078
2	1.059	13.232	58.310			
3	0.945	11.816	70.126			
4	0.828	10.350	80.477			
5	0.652	8.152	88.628			
6	0.403	5.040	93.668			
7	0.285	3.565	97.233			
8	0.221	2.767	100.000			

Extraction method Principal component analysis

Table 5.16 Analysis of anticipatory adjustment items

Anticipatory adjustment	Factor load	Mean	Standard deviation
AA10-Favorable housing conditions	0.654	3.7813	0.95609
AA11-No effect of changes back home	0.641	3.7625	0.91451
AA12-Smooth personal repatriate adjustment	0.586	4.1063	0.68769
AA9-Improved social status	0.561	3.7813	1.00734
AA5-Role clarity and hierarchy of job	0.543	3.6500	0.98543
AA4-Smooth family adjustment	0.494	3.8813	0.76374
AA6-Lesser role conflict	0.487	3.6500	0.91939
AA2-Repatriation expenses	0.441	3.7875	0.83468
Overall		3.8000	0.59763

Extraction method Principal component analysis

and fit between the Anticipatory Adjustment and the Repatriate Adjustment. The factor loading of each item indicates that the measurement scales of the constructs were stable and consistent.

Factor loadings range between 0.441 and 0.654 (Table 5.16) and significant at 0.05 levels, indicating high validity of the measurements and reflecting that our respondents could discriminate among all the measures, offering evidence of validity (Podsakoff & Organ, 1986). To determine the constructs' validity, the path coefficients (i.e., factor loadings) from the latent constructs to the corresponding indicators for all the constructs were analyzed (Sujan et al., 1994). The results in Table 5.12 show the factor loadings for each item in relation to the various factors and indicate that all the factor loadings are positive and all items are significantly loaded to the intended factors. Moreover, the results indicated that when the data will be subjected to factor analysis, there will be no or little multiple loading in the rotated factor loading table.

5.5.4 Respondents' Perception Regarding Reverse Culture Shock Items

The suitability of raw data for factor analysis obtained from respondents is examined through KMO value, Bartlett's test of Sphercity and *p*-value, (Table 5.10) indicating sufficient common variance and correlation matrix (Dess et al., 1997). The process of Principal Component Analysis (PCA) with Varimax Rotation brought the construct to the level of nine statements out of fifteen statements originally kept in the domain of access. The communality for nine items was above 0.5, indicating moderate to high degree of linear association among the variables. The factor loading ranges from 0.435 to 0.576 (Table 5.18), indicating that the factor loading values of the eight items of point towards relatively fair importance of items in defining the dimensions of Reverse Culture Shock. Table 5.17 indicates that the total cumulative variance explained by the factor Reverse Culture Shock experienced by Repatriates is 41.686%. The mean values varied between 2.2063 and 3.3188 and the total mean value stands at 2.77, which on a five point scale indicates the varying familiarity of the Repatriates to Reverse Culture Shock. The total computed standard deviation of the items stands at 0.730.

On the whole, all selected items significantly contribute towards this construct, thereby suggesting, the subjects believe that the congruity between Reverse Culture Shock and Repatriate Adjustment is important as the overall mean value of 2.7708 on a 5-point scale clearly indicates its relevance. This implies that most respondents perceived these nine statements as being of importance in general. The mean of the responses to Reverse Culture Shock factor item "Pre repatriation anxiety" is 0.472 and that of "Missing foreign culture" is 3.3188. The mean of Reverse Culture Shock dimensions with a possible range of Strongly Disagree "1 to 5" Strongly Agree, were 3.1563, 2.9750, 2.2063, 2.7688, 2.2375, 2.6563, 2.5688, 3.0500, 3.3188, respectively. The results point out that a majority of respondents "agree" that they are familiar with

Table 5.17 Total variance explained (reverse culture shock)

Component	Initial Eigen values			Extraction sums of squared loadings		
	Total	% of variance	Cumulative %	Total	% of variance	Cumulative %
1	3.752	41.686	41.686	3.752	41.686	41.686
2	1.527	16.968	58.654			
3	0.999	11.096	69.750			
4	0.769	8.545	78.295			
5	0.505	5.609	83.903			
6	0.465	5.161	89.065			
7	0.449	4.986	94.051			
8	0.293	3.259	97.310			
9	0.242	2.690	100.000			

Extraction method Principal component analysis

Table 5.18 Analysis of reverse culture shock items

Reverse culture shock	Factor load	Mean	Standard deviation
RCS2-Clash of expectations and reality	0.576	2.9750	1.18667
RCS6-Depression post repatriation	0.569	2.2375	0.98726
RCS7-Lonely and isolated post repatriation	0.537	2.6563	1.23902
RCS9-Changes in values and attitudes post repatriation	0.535	3.0500	1.10914
RCS11-Missing foreign culture	0.532	3.3188	1.03627
RCS8-Alienation post repatriation	0.511	2.5688	1.18493
RCS4-Sense of frustration, pre and post repatriation	0.486	2.7688	1.22973
RCS1-Pre-Repatriation anxiety	0.472	3.1563	1.13574
RS3-Problems in communication post repatriation	0.435	2.2063	1.11647
Overall		2.7708	0.73071

Extraction method Principal component analysis

the specified items of Anticipatory Adjustment. The mean values of items of Reverse Culture Shock are approximately equal. The standard deviation is 0.730, which means that the scores are spread away from the mean. Overall, the results indicated that the respondents viewed logical connection and fit between the Reverse Culture Shock and the Repatriate Adjustment. The factor loading of each item indicates that the measurement scales of the constructs were stable and consistent.

Factor loadings range between 0.435 and 0.576 (Table 5.18) and significant at 0.05 levels, indicating high validity of the measurements and reflecting that our respondents could discriminate among all the measures, offering evidence of validity (Podsakoff & Organ, 1986). To determine the constructs' validity, the path coefficients (i.e., factor loadings) from the latent constructs to the corresponding indicators for all the constructs were analyzed (Sujan et al., 1994). The results in Table 5.12 show the factor loadings for each item in relation to the various factors and indicate that all the factor loadings are positive and all items are significantly loaded to the intended factors. Moreover, the results indicated that when the data will be subjected to factor analysis, there will be no or little multiple loading in the rotated factor loading table.

5.6 Correlations

When the data is appropriate, it is possible to create a correlation matrix by calculating the correlations between each pair of variables. With respect to the correlation matrix, two things are important: the variables have to be inter-correlated, but they should not correlate too highly (extreme multi-co linearity and singularity) as this would cause difficulties in determining the unique contribution of the variables to a factor (Roberts et al., 2006).

In SPSS, now called PASW, the inter-correlation can be checked by using Bartlett's Test of spherecity, which "tests the null hypothesis that the original correlation matrix is an identity matrix" (Field, 2000). This test must be significant: when the correlation lattice is a identity matrix, there would be no correlations between the variables. Multi-co linearity then can be detected via the determinant of the correlation matrix, which can also be calculated in SPSS: if the determinant is greater than 0.00001, then there is no multi-co linearity. Correlation is used to measure the relation between two or more variables. Correlation coefficients can range from −1.00 to +1.00. The value of −1.00 represents a perfect negative correlation, while a value of +1.00 represents a perfect positive correlation. A value of 0.00 represents a lack of correlation.

The linear correlation presents Pearson correlation r between variables used in the questionnaire using 5% significance to support the questionnaire. A correlation test has been done with all variables impacting Repatriate adjustment back in the home country, as illustrated in the Table 5.19. The correlation matrix between various factors measured in the present study reveals that maximum correlation exists between Competency Transfer and Effective Repatriation (0.760) and correlation between Anticipatory Adjustment and Reverse Culture Shock is minimum (0.211) and negative. None of the correlations was higher than 0.8 which indicates that each variable can be distinct enough to measure different variables under the same construct (Sekaran, 2006).

The result shows that dependent variables are significantly correlated with all independent variables. Moreover, all variables are also significantly correlated with each other with only low to moderate coefficient scores. This low to moderate level of correlations between variables is normally expected to explain discriminant validity of the measures indicating the measurement scales' ability to discriminate between measures that are supposed to differ (Campbell & Fiske, 1959). It further anticipated that these correlations hint that the proposed theoretical framework will perform well in testing the theory.

Table 5.19 Correlation

	Competency transfer	Effective repatriation	Anticipatory adjustment	Reverse culture shock
Competency transfer	1			
Effective repatriation	0.760**	1		
Anticipatory adjustment	0.503**	0.549**	1	
Reverse culture shock	−0.222**	−0.343**	−0.211**	1

**Significant at 5% level of confidence

Table 5.19 shows that there is positive as well as negative correlation between the various constructs and thus, these factors make a good questionnaire for accessing the Antecedents of Repatriate Adjustment.

5.7 Regression Analysis

This section deals with the analysis of relationship between various constructs that were considered while conducting the study.

5.7.1 Impact Assessment of Demographic Variables on Competency Transfer

Multiple Regression analysis is done to examine the relationship between Competency Transfer as dependent variable and the demographic variables–Gender, Age, Marital status, Job profile, Length of stay overseas and Time elapsed since return from overseas as independent variables.

Table 5.20 indicates that 21.9% of variability in the dependent variable, Competency Transfer is accounted for by all of the six demographic variables together. The significant F-value at $p < 0.000$ indicates that this model is a good fit for the data and it means that at least one of the independent variables (demographic variables) is a significant predictor of the dependent variable Competency Transfer.

Based on Table 5.21, the regression equation is:

$$\begin{aligned}\text{Competency Transfer} = {} & 3.885 + 0.129(\text{Gender}) + 0.181(\text{Age}) \\ & + 0.103(\text{Marital status}) \\ & - 0.027(\text{Length of stay overseas}) \\ & - 0.355(\text{Job profile as Repatriate}) \\ & - 0.020(\text{Time elapsed since return to home country}).\end{aligned}$$

The β coefficients (Table 5.21) indicates that out of all six demographic variables, Age ($\beta = 0.181$; p-value $= 0.024$) and Job profile as a Repatriate ($\beta = -0.355$, and p-value $= 0.001$) has a significant influence on Competency Transfer. Whereas the variable Gender ($\beta = 0.129$; p-value $= 0.235$); Marital status ($\beta = 0.103$: p-value $=$ 0.376); Length of stay overseas ($\beta = -0.027$; p-value $= 0.604$) and Time elapsed since return to home country ($\beta = -0.020$; p-value $= 0.635$) have no significant impact on Competency Transfer.

Table 5.20 Regression analysis between demographic variables and competency transfer

Model	R	R square	Adjusted R square	Standard error of the estimate	Change statistics				
					R square change	F change	df1	df2	Sig. F change
1	0.468[a]	0.219	0.189	0.56713	0.219	7.164	6	153	0.000

[a]Cofficient

Table 5.21 Regression coefficients: demographic variables and competency transfer

Model		Un-standardized coefficients		Standardized coefficients	T	Sig
		B	Std. error	Beta		
1	(Constant)	3.885	0.445		8.732	0.000
	Gender	0.129	0.108	0.088	1.192	0.235
	Age	0.181	0.080	0.209	2.274	0.024
	Marital status	0.103	0.116	0.069	0.888	0.376
	Length of stay overseas	−0.027	0.052	−0.042	−0.519	0.604
	Job profile as repatriate	−0.355	0.103	−0.316	−3.429	0.001
	Time elapsed since return to home country	−0.020	0.042	−0.035	−0.476	0.635

Significant at $p < 0.05$

5.7.2 *Impact Assessment of Demographic Variables on Effective Repatriation*

Multiple Regression analysis is done to examine the relationship between Effective Repatriation as dependent variable and the demographic variables—Gender, Age, Marital status, Job profile, Length of stay overseas and Time elapsed since return from overseas, as independent variables.

Table 5.22 indicates that 14.1% of variability in the dependent variable, Effective Repatriation is accounted for by all of the six demographic variables together. The significant F-value at $p < 0.001$ indicates that this model is a good fit for the data and it means that at least one of the independent variables (demographic variables) is a significant predictor of the dependent variable Effective Repatriation.

Based on Table 5.23, the regression equation is:

$$\begin{aligned}\text{EFFECTIVE REPATRIATION} = {} & 3.678 + 0.160(\text{Gender}) + 0.160(\text{Age}) \\ & + 0.061(\text{Marital status}) \\ & - 0.036(\text{Length of stay overseas}) \\ & - 0.250(\text{Job profile as Repatriate}) \\ & - 0.051(\text{Time elapsed since return to home country}).\end{aligned}$$

The β coefficients (Table 5.23) indicates that out of all six demographic variables, the variables Age (β = 0.160; *p*-value = 0.044) and Job profile as a Repatriate ($\beta =$ 0.250, and *p*-value = 0.015) have a significant influence on Effective Repatriation. Whereas the variables Gender (β = 0.113; *p*-value = 0.291); Marital status (β = 0.061: *p*-value = 0.597); Length of stay overseas (β = −0.036; *p*-value = 0.491) and

Table 5.22 Regression analysis between demographic variables and effective repatriation

Model	R	R square	Adjusted R square	Standard error of the estimate	Change statistics				
					R square change	F change	df1	df2	Sig. F change
1	0.375[a]	0.141	0.107	0.56011	0.141	4.175	6	153	0.001

[a]Cofficient

Table 5.23 Regression coefficients: demographic variables and effective repatriation

Model		Un-standardized coefficients		Standardized coefficients	T	Sig
		B	Std. Error	Beta		
1	(Constant)	3.678	0.439		8.370	0.000
	Gender	0.113	0.107	0.082	1.060	0.291
	Age	0.160	0.079	0.196	2.030	0.044
	Marital status	0.061	0.115	0.043	0.529	0.597
	Length of stay overseas	−0.036	0.052	−0.059	−0.691	0.491
	Job profile as repatriate	−0.250	0.102	−0.237	−2.452	0.015
	Time elapsed since return to home country	−0.051	0.042	−0.093	−1.211	0.228

Significant at $p < 0.05$

Time elapsed since return to home country ($\beta = -0.051$; p-value $= 0.228$) have no significant impact on Effective Repatriation.

5.7.3 *Impact Assessment of Demographic Variables on Anticipatory Adjustment*

Multiple Regression analysis is done to examine the relationship between Anticipatory Adjustment as dependent variable and the demographic variables—Gender, Age, Marital status, Job profile, Length of stay overseas and Time elapsed since return from overseas, as independent variables.

Table 5.24 indicates that 16.7% of variability in the dependent variable, Anticipatory Adjustment is accounted for by all of the six demographic variables together. The significant F-value at $p < 0.000$ indicates that this model is a good fit for the data and it means that at least one of the independent variables (demographic variables) is a significant predictor of the dependent variable Anticipatory Adjustment.

Based on Table 5.25, the regression equation is:

$$\begin{aligned}\text{Anticipatory Adjustment} = {} & 3.155 + 0.123(\text{Gender}) + 0.242(\text{Age}) \\ & + 0.104(\text{Marital status}) \\ & - 0.013(\text{Length of stay overseas}) \\ & - 0.017(\text{Job profile as Repatriate}) \\ & - 0.023(\text{Time elapsed since return to home country}).\end{aligned}$$

Table 5.24 Regression analysis between demographic variables and anticipatory adjustment

Model	R	R square	Adjusted R square	Standard error of the estimate	Change statistics				
					R square change	F change	df1	df2	Sig. F change
1	0.409[a]	0.167	0.135	0.44040	0.167	5.125	6	153	0.000

[a]Cofficient

Table 5.25 Regression coefficients: demographic variables and anticipatory adjustment

Model		Un-standardized coefficients		Standardized coefficients	T	Sig
		B	Std. error	Beta		
1	(Constant)	3.155	0.345		9.133	0.000
	Gender	0.123	0.084	0.110	1.457	0.147
	Age	0.242	0.062	0.372	3.917	0.000
	Marital status	0.104	0.090	0.092	1.150	0.252
	Length of stay overseas	−0.013	0.041	−0.027	−0.319	0.750
	Job profile as repatriate	−0.017	0.080	−0.021	−0.216	0.829
	Time elapsed since return to home country	−0.023	0.033	−0.053	−0.699	0.485

Significant at $p < 0.001$

The β coefficients (Table 5.23) indicate that out of all six demographic variables, only one variable i.e. Age (β = 0.242; *p*-value = 0.000) has a significant influence on Anticipatory Adjustment, whereas Gender (β = 0.123; *p*-value = 0.147); Marital status (β = 0.104; *p*-value = 0.252); Length of stay overseas (β = −0.013; *p*-value = 0.750); Job profile as a Repatriate (β = −0.017, and *p*-value = 0.829); and Time elapsed since return to home country (β = −0.023; *p*-value = 0.485) have no significant impact on Anticipatory Adjustment.

5.7.4 *Impact Assessment of Demographic Variables on Reverse Culture Shock*

Regression analysis is done to examine the relationship between Reverse Culture Shock as dependent variable and the demographic variables—Gender, Age, Marital status, Job profile, Length of stay overseas and Time elapsed since return from overseas, as independent variables.

Table 5.26 indicates that 11.1% of variability in the dependent variable, Reverse Culture Shock is accounted for by all of the six demographic variables together. The significant F-value at $p < 0.00$ indicates that this model is a good fit for the data and it means that at least one of the independent variables (demographic variables) is a significant predictor of the dependent variable Reverse Culture Shock.

Based on Table 5.27, the regression equation is:

Table 5.26 Regression analysis between demographic variables and reverse culture shock

Model	R	R square	Adjusted R square	Standard error of the estimate	Change statistics				
					R square change	F change	df1	df2	Sig. F change
1	0.333[a]	0.111	0.076	0.52752	0.111	3.172	6	153	0.001

[a]Cofficient

Table 5.27 Regression coefficients: demographic variables and reverse culture shock

Model		Un-standardized coefficients		Standardized coefficients	T	Sig
		B	Std. Error	Beta		
1	(Constant)	2.345	0.414		5.668	0.000
	Gender	0.179	0.101	0.139	1.776	0.078
	Age	−0.091	0.074	−0.121	−1.232	0.220
	Marital status	−0.062	0.108	−0.048	−0.577	0.565
	Length of stay overseas	0.115	0.049	0.204	2.353	0.020
	Job profile as repatriate	0.072	0.096	0.074	0.750	0.454
	Time elapsed since return to home country	0.122	0.039	0.242	3.079	0.002

Significant at $p < 0.05$

$$
\begin{aligned}
\text{Reverse Culture Shock} = {} & 2.345 + 0.179(\text{Gender}) - 0.091(\text{Age}) \\
& - 0.062(\text{Marital status}) \\
& + 0.115(\text{Length of stay overseas}) \\
& - 0.072(\text{Job profile as Repatriate}) \\
& - 0.122(\text{Time elapsed since return to home country}).
\end{aligned}
$$

The β coefficients (Table 5.27) indicate that out of all the six demographic variables, Length of stay overseas (β = 0.115; *p*-value = 0.020) and Time elapsed since return to the home country (β = 0.122, and *p*-value = 0.002) have a significant influence on Reverse Culture Shock). Whereas the variable Age (β = −0.091; *p*-value = 0.220); (Gender (β = 0.179; *p*-value = 0.078); Marital status (β = −0.062: *p*-value = 0.565); Job profile as a repatriate (β = 0.072; *p*-value = 0.454) have no significant impact on Reverse Culture Shock.

5.8 Hypotheses Testing

H1: Reverse Culture Shock Impacts Anticipatory Repatriation Adjustment

To test for the first hypothesis, we proceeded with Exploratory Factor Analysis to converge at two sub factors of Reverse Culture Shock (independent variable). Correlation between the factor groups of Reverse Culture Shock and average means of data from variables comprising Anticipatory Repatriate Adjustment was done, followed by Multiple Regression analysis between the identified factor groups of Reverse Culture Shock and the average mean of the statements of Anticipatory

Repatriate Adjustment, to determine the plausible effect of Reverse Culture Shock on Anticipatory Adjustment.

A. Exploratory Factor analysis to identify Factor Groups underlying Reverse Culture Shock

Factor analysis is a data reduction technique which is used in identifying a smaller number of factors underlying a large number of observed variables. Variables having a high correlation between them are to a great extent autonomous of different subsets of variables and are thus combined into factors. After already having done the factor analysis of the statements comprising the four constructs and ensuring the reliability of the statements by KMO and Bartlett's Test (Table 5.10), the technique of Exploratory Factor Analysis has been used through SPSS (17 version) with Principal Component Analysis along with Varimax rotation for summarization of the total data falling under the construct of "Reverse Culture Shock", measured through the research instrument (Questionnaire), to arrive at meaningful factor groups. Such grouping of the statements of variables pertaining to the measurement of Reverse Culture Shock, in distinct areas of importance, yielded two factor groups; "Alienation and Segregation" and "Loneliness and Anxiety", covering different dimensions of Reverse Culture Shock experienced by Repatriates. The implications for these factors are in relevance to the importance of these factors in impacting Anticipatory Repatriate Adjustment. Subsequently, multiple regression analysis between the RCS factor groups and ANOVA was explored and discussed.

Factor Analysis results divided the nine statements of the independent construct, Reverse Culture Shock (RCS), into two factor groups; "Alienation and Segregation" and "Loneliness and Anxiety".

Factor 1, labeled as "Alienation and Segregation" comprises of four statements such as: RCS2 (Clash of expectations), RCS9 (Changes in values and attitudes, post-repatriation), RCS8 (Alienation post repatriation) and RCS3 (Problems in communication post- repatriation). The item with the highest factor loading is "clash of expectations" (0.576), indicating high association with this factor. The item "Problems in communication post repatriation" has the lowest factor loading (0.435), indicating least association with the factor (Table 5.28).

Factor 2, labeled as "Loneliness and Anxiety" envisages five statements such as: RCS6 (Depression post repatriation), RCS7 (Lonely and isolated post repatriation), RCS11 (Missing foreign culture), RCS4 (Sense of frustration, pre and post repatriation) and RCS1 (Pre-repatriation anxiety). The item with the highest factor loading is "Depression post-repatriation" (0.569), indicating high association with the factor. Item "Pre repatriation anxiety" has the lowest factor loading (0.472), indicating low association with this factor (Table 5.28).

The mean values and factor loadings of the statements pertaining to the two factor groups of the construct "Reverse Culture Shock" are summarized in Table 5.28.

Table 5.29 depicts the Descriptive Statistics of the two factor groups of Reverse Culture Shock (Independent variable) and the average of the means of Anticipatory Adjustment variables (Dependent variable), which comprises the mean and standard deviation of the aforementioned constructs.

Table 5.28 Factor groups underlying reverse culture shock

Factor	Variables	Mean	Factor loading
(1) Alienation and segregation	RCS2 (Clash of expectations)	2.9750	0.576
	RCS9 (Changes in values and attitudes, post repatriation)	3.0500	0.535
	RCS8 (Alienation post repatriation)	2.5688	0.511
	RCS3 (Problems in communication post repatriation)	2.2063	0.435
(2) Loneliness and anxiety	RCS6 (Depression post repatriation)	2.2375	0.569
	RCS7 (Lonely and isolated post repatriation)	2.6563	0.537
	RCS11 (Missing foreign culture)	3.3188	0.532
	RCS4 (Sense of frustration, pre and post repatriation)	2.7688	0.486
	RCS1 (Pre-repatriation anxiety)	3.1563	0.472

Table 5.29 Descriptive statistics (reverse culture shock factor groups and anticipatory adjustment)

	Mean	Std. deviation	N
AA avg	3.8398	0.47343	250
RCS Factor1	2.9375	0.87640	250
RCS Factor2	2.8275	0.79952	250

B. Correlation between Anticipatory Adjustment and the two Factor Groups of Reverse Culture Shock

Correlation is a bi-variate analysis used to know the association between two variables. Three types of correlations are generally applied in statistical analysis of data. Pearson Correlation is used in the present study to find out the degree to which there is a linear association between variables. The value of Pearson r correlation lies between -1 to $+1$ from perfect positive to perfect negative linear association (Malhotra & Dash, 2010).

Here RCS = Reverse Culture Shock, RCS Factor 1 = Alienation and Segregation, RCS Factor 2 = Loneliness and Anxiety, AA avg = Average of the means of variables defining Anticipatory Repatriate Adjustment. A correlation test was done between the said variables. The results (Table 5.30), show negative but significant correlation between RCS Factor 1 and AA avg (-0.119) and negative but significant correlation between RCS Factor 2 and AA avg (-0.334). Therefore the correlations for mean scale scores in Table 5.30 indicates that the dependent variable Anticipatory Adjustment and the independent variable Reverse Culture Shock are moderately correlated as the correlation coefficients are less than 0.5 ($p < 0.5$). The results of

Table 5.30 Correlations

		AA avg	RCS factor1	RCS factor2
Pearson correlation	AA avg	1.000		
	RCS factor1	−0.119	1.000	
	RCS factor2	−0.334	0.613	1.000

*Significant at 0.05 level of significance

the correlation matrix reveal that there exists a correlation among the variables of Reverse Culture Shock and those of Anticipatory Adjustment.

Given the significance of correlation between the respective factors of Reverse Culture Shock and Anticipatory Adjustment (Table 5.30), multiple regression analysis was applied, whose conclusion is shown in Table 5.31.

Multiple Regression analysis has been applied between identified factors groups of Reverse Culture Shock, experienced by Repatriates and average mean of the variables defining Anticipatory Adjustment in the study. The two identified factors of RCS are Alienation and segregation (F1) and Loneliness and Anxiety (F2) and have been used as independent variables while the average mean of the statements of Anticipatory Adjustment (AA) has been taken as the dependent variable (Table 5.30).

The multiple R value of 0.351 indicated a significant relationship between the variables and the adjusted R squared (0.112) suggests that the Reverse Culture Shock factors account for 11.2% of the variance in Anticipatory Repatriate Adjustment. $F = 11.050$; $p < 0.000$ (Table 5.32) after applying ANOVA wherein F Change is equal to 11.050. The results indicate that for each unit increase in independent variable, there is acceleration by 0.123 units (R-square = 0.123) and that the results are statistically significant, as depicted by β and t-values (Table 5.33).

Factor Groups are indicated by their serial number as in Table 5.28.

The standardized beta estimates for the factors of independent variable Reverse Culture Shock (Table 5.33) were significantly lesser than zero ($\beta = 0.074$ and −0.248, $p < 0.05$). The significant relation between dependent variable, Anticipatory Adjustment and independent variable, Reverse Culture Shock ($R = 0.351, p < 0.01$; $\beta = \beta = 0.074$ and −0.248, $p < 0.05$) (Tables 5.31, 5.33), demonstrate support for the fact that the factor has a significant relationship with Anticipatory Adjustment, albeit inversely.

The strength of the relationship between Reverse Culture Shock and Anticipatory Adjustment of Repatriates is represented in form of a regression equation, which is as follows:

$$\text{AA(Anticipatory Adjustment)} = \alpha + \beta 1\ F1 + \beta 2\ F2$$

$$\text{AA(Anticipatory Adjustment)} = 4.323 + 0.074(\text{RCS } F1) - 0.248(\text{RCS } F2)$$

where α is constant, β1 and β2 represent coefficients of regression for the independent factors of "Alienation and segregation" and "Loneliness and anxiety", respectively.

Table 5.31 Impact assessment of reverse culture shock on anticipatory repatriate adjustment

Model	R	R square	Adjusted R square	Standard error of the estimate	Change statistics				
					R square change	F change	df1	df2	Sig. F change
1	0.351	0.123	0.112	0.44608	0.123	11.050	2	157	0.000

(a) Predictors (Constant) RCS Factor 2, RCS Factor 1
(b) Dependent Variable AA avg

Table 5.32 ANOVA

Model		Sum of squares	df	Mean square	F	Sig
1	Regression	4.398	2	2.199	11.050	0.000[a]
	Residual	31.241	157	0.199		
	Total	35.638	159			

(a) Predictors (Constant), RCSfactor2, RCSfactor1
(b) Dependent Variable AA avg

Table 5.33 Regression coefficients

Model		Un-standardized coefficients		Standardized coefficients	t	Sig
		B	Std. error	Beta		
1	(Constant)	4.323	0.140		30.793	0.000
	RCS factor 1	0.074	0.051	0.137	1.446	0.150
	RCS factor 2	−0.248	0.056	−0.418	−4.421	0.000

Dependent variable AA avg

The results clearly indicate that the two factor groups of Reverse Culture Shock are significant in explaining 11.2% of variability in Anticipatory Adjustment, as the value of adjusted R-Square is 0.112 (Table 5.31). 7.4% of variability in Anticipatory Adjustment (AA) is accounted for by the first factor group RCS F1. Also 24.8% of variability in Anticipatory Adjustment (AA) is accounted for by the second factor group RCS F2 (Table 5.33). Thus the above regression equation depicts that Reverse Culture Shock impacts Anticipatory Adjustment.

The statistical results demonstrate support for the fact that factors Of Reverse Culture Shock (RCS) have an impact on Anticipatory Repatriate Adjustment (AA). This implies that the findings considerably denote that the criterion of the variables defining Anticipatory Adjustment, rely heavily on the construct of Reverse Culture Shock and that the latter has an impact on the former whereby the Anticipatory Adjustment while overseas with respect to the foreseen adjustment in home country, gets negatively impacted, thereby creating a gap between anticipated expectations of Repatriation outcomes and the reality of Repatriation back to home country. This indicates that the impediments of Reverse Culture Shock are significantly predictive of an unfavorable impact on Anticipatory repatriation expectations of a repatriate and thereby hamper the process of smooth repatriation adjustment in home country. Effective management of Repatriation via Repatriation assistance and elimination of Reverse Culture Shock factors was significantly predictive of a favorable Repatriate adjustment in Job, Organizational, interaction and non-work areas, once the repatriate is back home. These findings also validate the proposition P1 (Black et al., 1991) that each facet of anticipatory expectations will have its strongest impact on its corresponding facet of in-country repatriation adjustment.

Therefore this validates the Hypothesis H1 that Reverse Culture Shock impacts Anticipatory Adjustment of Repatriates.

H2: Competency Transfer is significantly a function of Effective Repatriation

To test for the second hypothesis, we proceeded with exploratory factor analysis to converge at two sub factors of Effective Repatriation (independent variable). Correlation between the factor groups of Effective Repatriation and average means of data from variables comprising Competency Transfer of Repatriates was done, followed by Multiple Regression analysis between the identified factor groups of Effective Repatriation and the average mean of the statements of Competency Transfer, to determine the effect of Effective Repatriation predictors on Competency Transfer outcomes.

A. Exploratory Factor Analysis to identify Factor Groups underlying Effective Repatriation

Factor analysis is a data reduction technique which is used in identifying a smaller number of factors underlying a large number of observed variables. Variables having a high correlation between them are to a great extent autonomous of different subsets of variables and are thus combined into factors. After already having done the factor analysis of the statements comprising the four constructs and ensuring the reliability of the statements by KMO and Bartlett's Test (Table 5.10), the technique of Exploratory Factor Analysis has been used through SPSS (17 version) with Principal Component Analysis along with Varimax rotation for summarization of the total data falling under the construct of "Effective Repatriation", measured through the research instrument (Questionnaire), to arrive at meaningful factor groups. Such grouping of the statements of variables pertaining to the measurement of Reverse Culture Shock, in distinct areas of importance, yielded two factor groups; "Career Description, Progression, Training Role and Social Status, Post-Repatriation" and "Acquired Acumen and Job Growth, Role Of Communication, Changes Back Home and Repatriation Stress", covering different dimensions of Effective Repatriation (ER) of Repatriates. The implications for these factors are in relevance to the importance of these factors in Repatriate Competency Transfer. Subsequently, multiple regression analysis between the ER factor groups and ANOVA was explored and discussed.

Factor Analysis results divided the nine statements of the independent construct, Effective Repatriation (ER) of Repatriates into two factor groups:

Factor 1, labeled as "Career Aspects" consists of five statements: ER3 (Job description), ER2 (Job position and hierarchy), ER4 (Career progression ahead of colleagues), ER7 (Social status post-repatriation) and ER1 (Repatriation training and benefit). The item with the highest factor loading is "Job description" (0.687), indicating high association with this factor. The item "Repatriation training and benefit" has the lowest factor loading (0.436), indicating least association with the factor (Table 5.34).

Factor 2, labeled as "Acumen, Communication, Changes and Repatriation Stress" envisages four statements such as: ER8 (Acquired acumen and career development),

Table 5.34 Factor groups underlying, effective repatriation

Factor	Variables	Mean	Factor loading
(1) Career aspects	ER3 (Job description)	3.4625	0.687
	ER2 (Job position and hierarchy)	3.4438	0.591
	ER4 (Career progression ahead of colleagues)	3.4125	0.512
	ER7 (Social status post-repatriation)	3.0438	0.441
	ER1 (Repatriation training and benefit)	3.3688	0.436
(2) Acumen, communication, changes and repatriation stress	ER8 (Acquired acumen and career development)	3.7688	0.539
	ER11 (Effect of changes in home country)	3.4063	0.514
	ER9 (No repatriation stress)	3.1313	0.495
	ER5 (Role of communication in uncertainty elimination)	3.6875	0.479

ER11 (Effect of changes in home country), ER9 (No repatriation stress) and ER5 (Role of communication in uncertainty elimination). The item with the highest factor loading is "Acquired acumen and career development" (0.539), indicating high association with the factor. Item "Role of communication in uncertainty elimination" has the lowest factor loading (0.479), indicating low association with this factor (Table 5.34).

The mean values and factor loadings of the statements pertaining to the two factor groups of the construct "Effective Repatriation" are summarized in Table 5.34.

Table 5.35 depicts the Descriptive Statistics of the two factor groups of Effective Repatriation (Independent variable) and the average of the means of Competency Transfer variables (dependent variable), which comprises of the mean and standard deviation of the aforementioned constructs.

B. Correlation between Competency Transfer and two Factor Groups of Effective Repatriation

Here ER = Effective Repatriation, ER Factor 1 = Career Description, Progression, Training role and Social Status post-Repatriation, ER Factor 2 = Acquired acumen and Job growth, Role of communication, Changes back home and Repatriation stress, CT avg = Average of the means of variables defining Competency

Table 5.35 Descriptive statistics (effective repatriation factor groups and competency transfer)

	Mean	Std. Deviation	N
CTavg	3.7512	0.62965	250
ER factor1	3.3463	0.78851	250
ER factor2	3.4984	0.80997	250

Table 5.36 Correlations

		CTavg	ER factor1	ER factor2
Pearson correlation	CTavg	1.000		
	ER factor1	0.655*	1.000	
	ER factor2	0.703*	0.645*	1.000

*Significant at 0.05 level of significance

Transfer. A correlation test was done between the said variables. The results (Table 5.36), show a positive and significant correlation between ER Factor 1 and CT avg (0.655) and negative but significant correlation between ER Factor 2 and CT avg (0.645). Therefore the correlations for mean scale scores in Table 5.36 indicates that the dependent variable Competency Transfer and the independent variable Effective Repatriation are moderately correlated as the correlation coefficients are less than 0.5 ($p < 0.5$).

The results of the correlation matrix reveal that there is a positive correlation among the variables of Effective Repatriation and those of Competency Transfer.

Given the significance of this correlation, multiple regression analysis was done, between independent variable, identified as factor groups of Effective Repatriation (ER), of Repatriates and their Competency Transfer (dependent variable), whose conclusion is shown in Table 5.37.

The two identified factors of ER are "Career Aspects" and "Acumen, Communication, Changes and Repatriation Stress", and have been used as independent variables while the average mean of the statements of Competency Transfer (CT) has been taken as the dependent variable (Table 5.34).

The results indicate that for one unit increase in independent variable, there is an acceleration by 0.564 units (R Square = 0.564) and that the results are statistically significant as depicted by β and t-values. The multiple R value of 0.751 indicates a significant relationship between the variables and the adjusted R square (0.558) suggests that the Effective Repatriation factors account for 55.8% of the variance in Competency Transfer. $F = 101.394$; $p < 0.000$ (Table 5.38) and Table 5.37 summarize these findings. Factor Groups are indicated by their serial number as in Table 5.34.

The standardized beta estimates for the dependent variable Competency Transfer (Table 5.39) were significantly lesser than zero ($\beta = 0.276$ and 0.373, $p < 0.05$). The statistical results demonstrate support for the fact that the Competency Transfer is significantly a function of Effective Repatriation.

The strength of the relationship between Effective Repatriation and Competency Transfer is represented in form of a regression equation, which is as follows:

$$\text{CT (Competency Transfer)} = \alpha + \beta 1\ F1 + \beta 2\ F2$$

$$\text{CT (Competency Transfer)} = 1.523 + 0.276(\text{ER } F1) + 0.373(\text{ER } F2)$$

where α is constant, β1 and β2 represent coefficients of regression for the independent factors of Effective Repatriation. As observed from the findings in Table 5.37, the R

Table 5.37 Impact assessment of effective repatriation on competency transfer

Model	R	R square	Adjusted R square	Standard error of the estimate	Change statistics				
					R square change	F change	df1	df2	Sig. F change
1	0.751	0.564	0.558	0.41857	0.564	101.394	2	157	0.000

(a) Predictors (Constant) ER Factor 2, ER Factor 1
(b) Dependent Variable CT avg

Table 5.38 ANOVA

Model		Sum of squares	df	Mean square	F	Sig
1	Regression	35.529	2	17.765	101.394	0.000[a]
	Residual	27.507	157	0.175		
	Total	63.036	159			

(a) Predictors (Constant), ER Factor2, ER Factor1
(b) Dependent Variable CTavg

Table 5.39 Regression coefficients

Model		Un-standardized coefficients		Standardized coefficients	t	Sig
		B	Std. error	Beta		
1	(Constant)	1.523	0.160		9.504	0.000
	ER factor 1	0.276	0.055	0.346	5.020	0.000
	ER factor 2	0.373	0.054	0.479	6.948	0.000

value of 0.751 indicates a significant positive relationship between the variables and the adjusted R square (0.558) suggesting that the Effective Repatriation accounts for 55.8.5% of the variance in Competency Transfer ($F = 101.394$; $p < 0.000$) as shown in Table 5.38, after applying ANOVA. F change is equal to 101.394 (Table 5.38). The results clearly indicate that the two factors of Effective Repatriation are significant in explaining 55.8% of Competency Transfer, as the value of adjusted R- Square is 0.558.

27.6% variability in CT (Competency Transfer), the dependent variable is accounted for by the first factor group variables i.e. ER F1. Also 37.3% of variability in CT is accounted for by the second factor group i.e. ER F2.

Thus the above regression equation depicts that Competency Transfer is significantly a function of Effective Repatriation. The statistical results demonstrate support for the fact that factors of Effective Repatriation (ER) have an impact on Competency Transfer (CT). This implies that the findings considerably denote that the criterion of the variables defining Competency Transfer, rely heavily on the construct of Effective Repatriation and that the latter has an impact on the former whereby Effective Repatriation considerably paves way for the transfer of the competencies of a Repatriate. Therefore this validates the Hypothesis H2; Competency Transfer is significantly a function of Effective Repatriation.

5.9 Comparative Analysis (Analysis of Variance)

5.9.1 Anova on the Basis of Gender

See the Table 5.40.

Table 5.41 shows output from one-way ANOVA using the demographic variable of Gender subdivided into male and female categories. The findings suggest that the "*p*" value is more than 0.05 indicating no significant mean difference exist in the nature of Repatriate adjustment in all the constructs with regard to Gender.

Table 5.40 Gender wise analysis

Constructs	Male = 205		Female = 45	
	Mean	Standard deviation	Mean	Standard deviation
Competency transfer	3.7480	0.62869	3.7613	0.64106
Effective repatriation	3.5057	0.57007	3.5445	0.66762
Anticipatory adjustment	3.8284	0.47038	3.8763	0.48769
Reverse culture shock	2.9838	0.53878	3.1347	0.57157

Table 5.41 ANOVA gender wise analysis

Constructs		Sum of squares	df	Mean square	F	Sig	Remarks
Competency transfer	Between groups	0.005	1	0.005	0.013	0.910	Insignificant
	Within groups	63.031	158	0.399			
	Total	63.036	159				
Effective repatriation	Between groups	0.044	1	0.044	0.124	0.726	Insignificant
	Within groups	55.814	158	0.353			
	Total	55.857	159				
Anticipatory adjustment	Between groups	0.066	1	0.066	0.295	0.588	Insignificant
	Within groups	35.572	158	0.225			
	Total	35.638	159				
Reverse culture shock	Between groups	0.660	1	0.660	2.210	0.139	Insignificant
	Within groups	47.212	158	0.299			
	Total	47.872	159				

Table 5.42 Age wise analysis

Constructs	20–29 yrs N = 69		30–39 yrs N = 105		40–49 yrs N = 65		50–59 yrs N = 11	
	Mean	S.D	Mean	S.D	Mean	S.D	Mean	S.D
Competency transfer	3.5619	0.5968	3.6356	0.5768	4.2656	0.5259	4.1700	0.5359
Effective repatriation	3.3364	0.5404	3.4796	0.5942	3.8306	0.5528	3.9200	0.5423
Anticipatory adjustment	3.6464	0.4754	3.8091	0.3993	4.1706	0.4719	4.0800	0.4612
Reverse culture shock	3.1060	0.4976	2.9815	0.5797	2.9962	0.5514	2.7300	0.5513

where S.D. = Standard Deviation

As evident from the table, there exist no significant difference in response to various constructs, between male and female respondents, as value of $p > 0.05$ level of significance. So, on the basis of these findings, we can say that in both the hypotheses, Gender plays no significant in determining the constructs of the said hypotheses. The mean values, as suggested in Table 5.40 don't show much variance across the various constructs in terms of Gender.

5.9.2 Anova on the Basis of Age

Table 5.43 depicts age-wise output from one way ANOVA using different constructs of Repatriate Adjustment, as researched in the study. Variance of group is not same, as far as the first three constructs are concerned as the value of "*p*" is less than 0.05, indicating significant mean difference exist among respondents of different age groups. Whereas in case of Reverse Culture Shock, variance of group is same as the value of "*p*" is more than 0.05, indicating insignificant mean difference exist among different age groups and therefore the variance across various group responses, as far as Reverse Culture Shock is concerned. This is amply clear from the mean value readings in Table 5.42, as far as the different group responses with respect to the four constructs are concerned.

5.9.3 Anova on the Basis of Marital Status

Table 5.45 indicates output on basis of marital status, from one way ANOVA using different constructs of Repatriation adjustment. Insignificant difference exists among respondents, as far as Effective Repatriation and Reverse Culture Shock are concerned, as value of "*p*" is more than 0.05 in case of these constructs and hence

Table 5.43 ANOVA age wise analysis

Constructs		Sum of squares	df	Mean square	F	Sig	Remarks
Competency transfer	Between groups	11.897	3	3.966	12.097	0.000	Significant
	Within groups	51.139	156	0.328			
	Total	63.036	159				
Effective repatriation	Between groups	5.147	3	1.716	5.278	0.002	Significant
	Within groups	50.710	156	0.325			
	Total	55.857	159				
Anticipatory adjustment	Between groups	5.610	3	1.870	9.714	0.000	Significant
	Within groups	30.029	156	0.192			
	Total	35.638	159				
Reverse culture shock	Between groups	0.566	3	0.189	0.622	0.602	Insignificant
	Within groups	47.306	156	0.303			
	Total	47.872	159				

the variance of the groups is same. As far as the constructs of Competency Transfer and Anticipatory Adjustment are concerned, variance of group is not same as the value of "*p*" is less than 0.05 indicating significant difference exist among respondents who belong to different categories of marital status. The mean values under the specific categories of the said demographic suggest the same wherein it shows a gradation as far as the constructs of Competency Transfer and Anticipatory Adjustment is concerned and remains nearly the same, as far as the constructs of Effective Repatriation and Reverse Culture Shock is concerned (Table 5.44).

5.9.4 Anova on the Basis of Length of Stay Overseas

Table 5.47 shows one way ANOVA with respect to various categories with respect to the four constructs, under the demographic, Length of stay overseas. Variance of group is same as the value of "*p*" is more than 0.05, indicating insignificant mean difference exist among respondents belonging to different categories with regard to different dimensions of all the four constructs of Repatriate Adjustment. This is also evident from the mean values, which remain more or less the same, under various categories demarcated under the said demographic (Table 5.46).

Table 5.44 Analysis on basis of marital status

Constructs	Single N = 54		Married N = 188		Others N = 08	
	Mean	Std. deviation	Mean	Std. deviation	Mean	Std. deviation
Competency transfer	3.4400	0.5548	3.8147	0.6213	3.8975	0.8894
Effective repatriation	3.2829	0.5110	3.5667	0.5917	3.4800	0.9092
Anticipatory adjustment	3.6007	0.4173	3.8904	0.4658	3.8950	0.6928
Reverse culture shock	3.1500	0.5144	2.9875	0.5482	3.1350	0.7955

Table 5.45 ANOVA on the basis of marital status

Constructs		Sum of squares	df	Mean square	F	Sig	Remarks
Competency transfer	Between groups	3.313	2	1.657	4.355	0.014	Significant
	Within groups	59.723	157	0.380			
	Total	63.036	159				
Effective repatriation	Between groups	1.856	2	0.928	2.698	0.070	Insignificant
	Within groups	54.001	157	0.344			
	Total	55.857	159				
Anticipatory adjustment	Between groups	1.940	2	0.970	4.520	0.012	Significant
	Within groups	33.698	157	0.215			
	Total	35.638	159				
Reverse culture shock	Between groups	0.661	2	0.331	1.100	0.336	Insignificant
	Within groups	47.211	157	0.301			
	Total	47.872	159				

5.9.5 Anova on the Basis of Job Profile

Tables 5.48 and 5.49 depict the results from one way ANOVA for different dimensions of the demographic, Job profile. For all the dimensions of the Job profile i.e. Low level, middle level and top level, the variance of group is not same as the value of "*p*" is less than 0.05, in case of the constructs of Competency Transfer, Effective Repatriation and Anticipatory Adjustment, indicating that a significant mean difference exist

Table 5.46 Analysis on basis of length of stay overseas

Constructs	< 6 months N = 64		7–12 months N = 85		13–24 months N = 58		> 25 N = 43	
	Mean	S.D	Mean	S.D	Mean	S.D	Mean	S.D
Competency transfer	3.6944	0.5069	3.6667	0.6582	3.7252	0.6166	3.9974	0.6463
Effective repatriation	3.4460	0.4935	3.4615	0.6614	3.5734	0.6956	3.5906	0.3037
Anticipatory adjustment	3.7992	0.2593	3.7895	0.4994	3.8880	0.5234	3.9016	0.4868
Reverse culture shock	2.8980	0.6233	2.9858	0.5314	3.0991	0.4800	3.0703	0.6104

where S.D. = Standard Deviation

Table 5.47 ANOVA on the basis of length of stay overseas

Constructs		Sum of squares	df	Mean square	F	Sig	Remarks
Competency transfer	Between groups	2.418	3	0.806	2.075	0.106	Insignificant
	Within groups	60.618	156	0.389			
	Total	63.036	159				
Effective repatriation	Between groups	0.618	3	0.206	0.582	0.628	Insignificant
	Within groups	55.239	156	0.354			
	Total	55.857	159				
Anticipatory adjustment	Between groups	0.413	3	0.138	0.610	0.609	Insignificant
	Within groups	35.225	156	0.226			
	Total	35.638	159				
Reverse culture shock	Between groups	0.796	3	0.265	0.879	0.453	Insignificant
	Within groups	47.077	156	0.302			
	Total	47.872	159				

among respondents under different Job profile categories. For all the dimensions of the Job profile i.e. Low level, middle level and top level, the variance of group is same as the value of "*p*" is more than 0.05, in case of the constructs of Reverse Culture Shock, indicating that an insignificant mean difference exists among respondents under the different categories, with respect to the construct of Reverse Culture Shock.

Table 5.48 Analysis on basis of job profile

Constructs	Top level N = 16		Middle level N = 174		Lower level N = 60	
	Mean	Std. deviation	Mean	Std. deviation	Mean	Std. deviation
Competency transfer	4.3504	0.52881	3.7034	0.57223	3.4004	0.59476
Effective repatriation	3.7858	0.57849	3.5490	0.57696	3.1204	0.48745
Anticipatory adjustment	3.9296	0.57500	3.8864	0.44548	3.5600	0.39675
Reverse culture shock	3.0558	0.56004	2.9693	0.57797	3.1992	0.35202

Table 5.49 ANOVA on the basis of job profile

Constructs		Sum of squares	df	Mean square	F	Sig	Remarks
Competency transfer	Between groups	12.069	2	6.035	18.589	0.000	Significant
	Within groups	50.967	157	0.325			
	Total	63.036	159				
Effective repatriation	Between groups	5.936	2	2.968	9.335	0.000	Significant
	Within groups	49.921	157	0.318			
	Total	55.857	159				
Anticipatory adjustment	Between groups	2.467	2	1.234	5.839	0.004	Significant
	Within groups	33.171	157	0.211			
	Total	35.638	159				
Reverse CULTURE SHOCK	Between groups	1.149	2	0.575	1.931	0.148	Insignificant
	Within groups	46.723	157	0.298			
	Total	47.872	159				

5.9.6 Anova on the Basis of Time Elapsed Since Return to Home Country

Tables 5.50 and 5.51 depicts the results from one way ANOVA for different dimensions of the demographic; Time elapsed since return to the home country. For all the dimensions of the said demographic i.e. less than six months, seven to twelve months, thirteen to eighteen months and more than nineteen months, the variance

Table 5.50 Analysis of time elapsed since return to home country

Constructs	< 6 months N = 47		7–12 months N = 73		13–18 months N = 62		> 19 months N = 68	
	Mean	S.D	Mean	S.D	Mean	S.D	Mean	S.D
Competency transfer	3.8893	0.5090	3.6475	0.7115	3.6777	0.4846	3.8349	0.6695
Effective repatriation	3.5513	0.3406	3.5498	0.7295	3.4660	0.5690	3.4804	0.5666
Anticipatory adjustment	3.7690	0.3248	3.8642	0.5588	3.9080	0.4004	3.8118	0.4932
Reverse culture shock	2.7793	0.7033	3.0144	0.4131	3.0920	0.4716	3.1380	0.5904

where S.D. = Standard Deviation

of group is same as the value of "*p*" is more than 0.05, in case of the constructs of Competency Transfer, Effective Repatriation and Anticipatory Adjustment, indicating that a insignificant mean difference exist among respondents under different categories. For all the dimensions of time elapsed since return to the home country the variance of group is not same as the value of "*p*" is less than 0.05, in case of the constructs of Reverse Culture Shock, indicating that a significant mean difference exists among respondents under the different categories, with respect to the construct of Reverse Culture Shock.

5.10 Conclusion

Basis the data analysis as elucidated above, it can be concluded that demographics play an important role in understanding the context of Repatriation management and its dependent variables. Moreover the variables of this study, namely, Competency Transfer, Effective Repatriation, Anticipatory Adjustment and Reverse Culture Shock, have contributed towards explaining the concept of Repatriation adjustment and the factors that come into play while the process of Repatriation of Corporate executives takes place fruitfully or otherwise. This analysis entails that there exists a major impact of Reverse Culture Shock on Anticipatory Adjustment of Repatriates and that the context of Competency Transfer and acumen optimization, is a consequence of Effective Repatriation. The congruity between the major constructs of the study in defining the broader concept of Repatriation adjustment has been accentuated and deliberated upon in an empirical manner with precision. The statistical significance of the results shown by using various tools further highlights that the study undertaken holds relevance for a plethora of domains under Repatriation management at large.

Table 5.51 ANOVA on the basis of time elapsed since return to the home country

Constructs		Sum of squares	df	Mean square	F	Sig	Remarks
Competency transfer	Between groups	1.642	3	0.547	1.391	0.248	Insignificant
	Within groups	61.394	156	0.394			
	Total	63.036	159				
Effective repatriation	Between groups	0.232	3	0.077	0.217	0.885	Insignificant
	Within groups	55.625	156	0.357			
	Total	55.857	159				
Anticipatory adjustment	Between groups	0.358	3	0.119	0.528	0.664	Insignificant
	Within groups	35.280	156	0.226			
	Total	35.638	159				
Reverse culture shock	Between groups	2.521	3	0.840	2.891	0.037	Significant
	Within groups	45.351	156	0.291			
	Total	47.872	159				

References

Anderson, J. C., & Gerbing, D. W. (1988). Structural equation modeling in practice: A review and recommended two-step approach. *Psychological Bulletin, 103*(3), 411–423.

Black, J. S., Gregersen, H. B., & Mendenhall, M. E. (1991). Toward a theoretical framework of repatriation adjustment. *Journal of International Business Studies, 23*(4), 737–760.

Burns, C., & Bush R. F. (1995). *Marketing research.* Prentice Hall.

Campbell, D. T., & Fiske, D. W. (1959). Convergent and discriminant validation by the multitrait-multimethod matrix. *Psychological Bulletin, 56*(2), 81–105.

DeCoster, J., (1998). *Overview of factor analysis.* Department of Psychology, University of Alabama. stat-help.com

Dess, G. G., Lumpkin, G. T., & Covin, J. G. (1997). Entrepreneurial strategy making and firm performance: Tests of contingency. *Strategic Management Journal, 18*(9), 677–695.

DeVellis,. (2003). A psychometric toolbox for testing validity and reliability. *Journal of Nursing Scholarship, 39*(2), 155–164.

Gliem, J. A., & Gliem, R. R. (2003). *Calculating, interpreting, and reporting Cronbach's alpha reliability coefficient for Likert-type scales.* Indiana University.

Malhotra, N. K., Agarwal, J., & Peterson, M. (1996). Methodological issues in cross-cultural marketing research: A state-of-the-art review. *International Marketing Review, 13*(5), 7–43.

Malhotra, N. K., & Dash, S. (2010). *Marketing research: An applied approach.* Pearson Education South Asia.

Newman, I., & Newman, C. (1994). *Conceptual statistics for beginners.* books.google.com.

Podsakoff, P. M., & Organ, D. W. (1986). Self-reports in organizational research: Problems and prospects. *Journal of Management, 12*(4), 531–544.

Roberts, P., Priest, H., & Traynor, M. (2006). Reliability and validity in research. *Nursing Standard, 20*(44), 5–41.

Runeson, P., & Höst, M. (2009). Guidelines for conducting and reporting case study research in software engineering. *Springer.com, 14*(2), 131–164.

Sekaran, U. (2006). *Research methods for business: A skill building approach*, 4Th edn. books.google.com.

Sujan, H., Weitz, B. A., & Kumar, N. (1994). Learning orientation, working smart, and effective selling. *Journal of Marketing, 58*(3), 39–52.

Zikmund, W. G., & Babin, B. J. (1997). *Exploring market research*. The Dryden Press.

Chapter 6
Summary and Conclusions

6.1 Introduction

The purpose of this chapter is to summarize the findings of the present research and recommend areas for further analysis. The first section of the chapter discusses the research objectives of the present study. A summary of the major results have been described. The second part of the chapter discusses practical and managerial implications of the research and proposes recommendations for future research.

Therefore this chapter presents a synoptic view of the study, summary of key findings which have been supplemented by conclusions and suggestions. The discussion that follows may provide useful inputs to Human resource departments in Corporate India, as well as to the Repatriates and researchers in the Repatriation domain. The conclusions drawn are based on the data analysis conducted, and have been presented in the preceding chapter.

6.2 Executive Summary

The relevance of Repatriate Management as a key strategic tool in an ever dynamic International Human Resource Management scenario in the corporate sector, as discussed in the earlier chapters, is not of relevance only for theoretical purposes but has immense significance in encouraging "Reverse brain drain" and retention of a wide pool of our demographic dividend. Huge expenses are incurred on corporate hiring, training and engagement of human resources, thereby making it a realm of significant value to retain and groom repatriates by facilitating their adjustment back in the home country. Ample focus has been given to the research and study of Expatriation. Of late significant research is being done on the domain of Repatriation which marks the completion of the whole cycle of global relocations and coming back home.

M. B. Nag and F. Ahmad Malik, *Repatriation Management and Competency Transfer in a Culturally Dynamic World*,
https://doi.org/10.1007/978-981-19-7350-5_6

The present study has been an attempt in this direction whereby the constructs of Competency Transfer, Anticipatory Adjustment, Reverse Culture Shock and Effective Repatriation have been dealt with in a detailed manner, to study Repatriation Adjustment and its relevance in the said aspects. The findings of the study convey that Repatriate adjustment is an immensely important aspect of retaining valuable talent in the Corporate sector and it would go a long way in harnessing human resources, if Repatriation is successful and Repatriate talent is optimized by the Human Resource Departments of organizations. The inherent cultural clauses and impediments of acculturation and Reverse Culture Shock would thereby be rendered harmless, if that be done more efficiently. A valuable outcome of such endeavors would lead to knowledge sharing and transfer, competency deliverance, less attrition of Repatriates and overall organizational growth and productivity.

6.3 Organization of the Study

The entire study is divided into six chapters.

Chapter I: Introduction

This chapter contains an overview of Repatriation Management in the context of a globalized world and from International Human Resource Management (IHRM) perspective, followed by the importance of Repatriate talent and its optimization, from an Indian business perspective.

The first chapter of this thesis serves as an introduction to the field of Repatriation Acculturation and Adjustment antecedents and formalization of what knowledge the research aims to contribute. The chapter starts with an insight into the context of the study area, followed by an overview of the Information Technology sector in India, while discussing the area of repatriation of executives in Corporate India and conceptualizing Repatriation Adjustment. The chapter follows discussion explaining the Research rationale and significance and an account of the structural formation of the thesis chapters. Repatriation as a concept has been deliberated upon in this chapter and its success as a strategic Human Resource Management tool for retaining manpower and talent has been highlighted, along with the various nuances which effect and channelize such talent in a particular direction.

Chapter II: Review of Literature

The second chapter focuses on the review of relevant literature. It contains an expanded discussion on the literature with a complete and comprehensive review of the various researches that have been carried out in the topics of interest for the current study. The purpose of this chapter is to build a theoretical framework for the research by reviewing existing literature. It provides a thorough review of literature related to the Repatriation Adjustment and management, as an intertwined concept relying heavily on associated parameters and domains of Anticipatory Adjustment, Effective Repatriation and acculturation with special focus on Reverse Culture Shock.

It also discusses the dimensions of efficient Expatriation which initiates the cycle of successful Repatriation wherein the sojourners' holistic interpretation as far as the Psychological, Cultural, Sociological, financial and various other correlated paradigms of cultural and organizational adjustment is concerned.

The chapter concentrates on the discussion of the main theoretical and empirical contributions and review of the existing literature in the field of corporate repatriation adjustment. The literature reviewed helps in problem definition and formulation of research hypothesis, by providing detailed insight into the research gap. The chapter follows the detailed discussion of the Theoretical Background, followed by explaining in detail the concepts of Effective Repatriation, barriers to repatriation and repatriate adjustment. A detailed account of the variables under study in context of the Repatriation is discussed, which includes reviewing the vast arena of analytical studies on expatriation adjustment, repatriation adjustment and management, performance, the opportunities and challenges associated with it in the realm of IHRM which calls for probing deeper into the aspect.

Chapter III: Sectored Profile and the Precursors Effecting Repatriation Adjustment

The Internationalization of business and global outsourcing of services have given rise to mammoth opportunities for international exposure for Indian corporate workforce, especially in the IT sector. Herein, multiple factors have discerning impact upon International assignments and organizations need to respect that simply prioritizing international business in terms of revenue would be incomplete without sound management policies, because as proven by research, human and financial costs of under execution/failure in the international business sector are extensively more extreme than in the domestic arena. Therefore, organizations can't underestimate the complex nature of human resource (HR) problems in the international arena.

India's software and services exports have been rising rapidly over the last decade and Indian IT professionals form a large chunk of the total knowledge workers working globally. Therefore, their motivations are largely guided by their perceptions of organizational support, both intrinsically challenging jobs and a certain degree of self management on the job. This in turn implies that HR professionals need to chart out tangible career paths, devise efficient performance management systems and challenging job designs so as to motivate employee satisfaction and performance, as an Expatriate and a Repatriate. Organizations need to pay attention to perceived Repatriation Concerns, so as to ensure that the individual does not perceive Repatriation to be a detriment to career development. It is pertinent to gather information from various sources in order to understand what needs to be done to improve and optimize Repatriate talent and its subsequent adjustment back home.

This chapter heavily emphasizes on the topic of Globalization in the International human resource management with expanded emphasis on Repatriate Management and Adjustment in the IT sector organizational context. It addresses the broad and cross-disciplinary concept of Sectoral profile and the precursors effecting Repatriation Adjustment, with focus on the dimension of Indian repatriate acculturation in multi faceted dimensions in the Information Technology Industry in India. Each construct is introduced and explained briefly in context of Repatriation of corporate executives, thereby providing firm background to the research hypothesis.

Chapter IV: Research Design and Methodology

This chapter illustrates the methodology used in this research, with details on the research design, the data collection stage, and the model selected for testing the hypotheses. It provides a detailed deliberation on the formulation of research hypotheses, constructed in order to achieve the research objectives. It accentuates the scope of the study, research strategy, sampling, questionnaire designing, pilot study, final data collection and tabulation.

Chapter V: Data Analysis and Interpretation

This chapter entails data analysis and interpretation of results derived from applying relevant statistical tools and techniques to the data. It reports the statistical analyses conducted on the data gathered during the experiment. This section provides the numerical data and details whether the findings were significant. Hypotheses testing and data analysis has been done by the using statistical techniques including Factor analysis, Multiple Regression, and ANOVA. Means, standard deviations along with correlation coefficients and reliability test results have been listed in detail in this chapter.

The chapter highlights Repatriation Adjustment and Competency Transfer interface, along with Anticipatory Adjustment and Reverse Culture Shock relationship, followed by empirical findings and an analysis of the significance of the relationship model developed for the research work. The findings statistically reveal the strength of model relationships shared among the model constructs.

Chapter VI: Summary, Conclusions and Recommendations

This chapter highlights the conclusions and suggestions emerging from the discussions. This chapter concludes the present research with a discussion of the meaning of the findings, research limitations, and recommendations for further research.

It concludes the present research work with a detailed summary discussion, limitations of the research and recommended areas for future research. The conclusion of the research also presents in the practical repatriation implications, in context of the Indian corporate.

6.4 Objectives of the Study and Their Achievement

On the basis of the hypotheses and the domain of the study, the following objectives of the research were set:

Objective 1: To study and evaluate the emerging phenomenon of Repatriation Management, in the context of IHRM

This objective was conceptualized to study the current expansion and internationalization of business organizations, in a scenario where organizations are rapidly moving from being purely domestic players to being cross-national players. Increased internationalization manifests itself in a set up where companies believe that if they do not have employees with global skills; the organizations will lose their competitiveness. Companies have been reassigning employees abroad during a long time of period. However, since the business world has turned out to be more global, there has likewise been an increment in expatriation and consequently the need for Effective Repatriation. To make it easier and minimize the risks of facing adjustment difficulties for employees, companies' Human Resource Departments, in particular, have an obligation to guarantee a smooth adjustment for employees coming back to home country.

While, a few researchers have inspected the issues with picking and preparing expatriates, few have examined the last portion of global assignments—the Repatriation Process. This primarily can be the reason that even though expatriation has been around quite some time, companies are still struggling with the repatriation part. The essential explanation behind repatriation process not seen as a vital issue may be a direct result of the perspective that the expatriate is just returning home and that this ought not to bring about any issues. However, studies have demonstrated that repatriating is frequently more difficult than expatriating. Research has additionally demonstrated that there are a few issues both practical and emotional, which are identified with the repatriation process.

The cost of failed international assignments is high, both financially for the organization and from an employee's perspective. Research proves that the cost of repatriation failure is manifold; one is "underutilization of key employees". Many repatriates believe that when they return they would be promoted, which is not always possible. As indicated by exploration, problems show up when the repatriates get back home and understand that they have lost ground in their profession, and that the associates that stayed in the domestic domain have better career growth. Much of the time, issues can likewise show up when the home organization neglects to recognize the new learning and acumen that the repatriate has obtained. Consequently, these employees are placed in lower positions in the organization than they in fact are appropriate for. In order to improve this problem companies must carefully consider the international expertise and knowledge that the repatriates acquires and moreover identify their definite value.

Research proves that the organizations that have poor repatriation processes will have employees that are unsatisfied with their overseas assignment. This will be communicated directly or indirectly to the rest of the employees, and can put off new personnel from going on international assignments. If the employees of an organization hesitate to go on international assignments, this can contribute to decreased cross-cultural knowledge and understanding. To make sure that new employees will go on overseas assignments, the organization must ensure that the repatriates are taken care of properly and that the employees recognize overseas assignments as an opportunity for their careers. They must ensure that their employees have a proper repatriation process. O'Sullivan (2002) characterized an Effective Repatriation move as, "The one in which, upon return, the repatriate: obtains entry to a job which perceives any recently procured global capabilities, encounters insignificant cross-society rearrangement difficulties; and reports low turnover expectations/intentions".

Researchers have also examined the factors affecting the repatriation process. Regardless of prior studies done in this field, there are relatively few proposals or adequate guidelines as to what is the best way to handle repatriation in an of Indian context in the IT sector, which exhibits a widespread phenomenon of Expatriation and Repatriation of employees, due to a global business scenario in the recent years. In the context of Information Technology (IT) services, Indian software firms provide IT solutions to organizations all around the globe.

Existing findings of international human resource practices recommend that customary parameters of Human Resource Management may demonstrate a lacking to manage portability needs or acknowledgment of the readiness of employees regarding global exchanges. The entire inquiry of expatriate management in multinational organizations includes a perplexing scope of issues. Research proposes that thorough performance management frameworks for expatriates are a long way from general. Also, problems in cultural adjustment which may have an effect on work performance ought to be considered while evaluating an expatriates' performance in a new role.

Consequently, it is believed that guidelines and more proposition on the most proficient method to effectively oversee repatriation could be created to improve the process. In the said context, it is conclusive to study Repatriation in corporate India within the domains of the constructs described in the study, namely Anticipatory Adjustment, Competency Transfer, Effect of Reverse Culture Shock and subsequently the credibility of Effective Repatriation. The findings of the research can be deliberated upon and analyzed to provide a base for a discussion about how and what practical solutions that companies could apply to meet the underlying needs in creation of an Effective Repatriation process. An experimental methodology was used to validate this objective by collection of data from NASSCOM accredited IT companies and subjecting it to various statistical tools, which proved that Repatriation concerns do effect the Expatriates returning back to home country.

The results indicated that increase in the effectiveness of Repatriation process is significantly predictive of overall Repatriate and organizational welfare. The summation of the findings of the two hypotheses, using Correlation and Multiple Regression analysis and one way ANOVA, confirm the said objective. Moreover theoretical interpretation of the same is elaborated upon in Chapter three.

Objective 2: To analyze the factors effecting Repatriate Adjustment in the home country

One of the essential hypothetical procedures identified with adjustment, whether it is repatriation adjustment, cross cultural adjustment, or adjustment after a domestic transfer, is reduction of uncertainty. Individuals entering a new environment are regularly questionable in the matter of what is worthy and unsatisfactory or proper and wrong. The adjustment procedure includes the reduction of that vulnerability, so as to realize what is satisfactory and practical. The aspect of "cultural shock" is to a great extent an arrangement of negative responses to experiences that confront an individual in a remote environment and of those schemata that they do not have a precise arrangement and comprehension of and which disable them from properly acting in the new circumstance. The objective was to investigate the factors effecting Repatriate adjustment in the home country. The four constructs with their respective items under study analyzed the said perspective in detail wherein the transfer of competencies as a result of Effective Repatriation techniques aided the adjustment antecedents of a Repatriate. Likewise, the negative impact of Reverse Culture Shock in the home country on repatriation, which may cause impediments in the fulfillment of Anticipatory Adjustment expectations while abroad, has been analyzed in detail, thereby holistically covering the process of Repatriate adjustment back in the home country.

It has been empirically analyzed in the study pertaining to the first hypothesis that Reverse Culture Shock in the home country does impact Anticipatory Adjustment expectations of Repatriates wherein, home-country organizations lose out in form of unsuccessful repatriation process involving costs for both the organization and the employee. The purpose of this study is to analyze the repatriation process of the IT sector employees in India and propose relevant recommendations, basis the findings. This is done by examining via the measurement scales of constructs, the most critical factors to achieve an Effective Repatriation process, pre departure to overseas, during stay and on return, pertaining to individual, job and non work factors. On statistical analysis of Effective Repatriation, anticipatory repatriation and Reverse Culture Shock and processing of the data collected basis it, it was observed that interplay of the said constructs resulted in transfer or failure of transfer of competencies of repatriates or ineffective repatriate adjustment. The demographics at play while collecting data and measuring the reliability and validity of constructs, indicated the respective outcomes. Likewise, the hypothesis validity pertaining to the significance of Reverse Culture Shock being an in country determinant of adjustment and its impact on Anticipatory Adjustment while abroad, indicates the various paradigms that go into the making of the said construct via the items it entails. Competency Transfer being a significant indicator of home country adjustment, when Effective Repatriation takes place.

The study validates the constructs given by Black et al. pertaining to home country Repatriate adjustment determinants wherein individual need for control and self-efficacy is positively related to all facets of in-country repatriation adjustment. Role clarity and role discretion is positively related to repatriation work adjustment and role conflict will be negatively related to repatriation work adjustment. Post-return training and orientation of sufficient rigor and content have a positive impact on all three facets of in-country repatriation adjustment. Congruent and clear organizational and individual repatriation career objectives and repatriation policies are positively related to in-country repatriation work adjustment. A downward shift in social status and poorer housing conditions are negatively related to all facets of repatriation adjustment.

Given the fact that the current study validates the above given benchmarks, it is imperative that international HR practices be designed so as to enable expatriate satisfaction which in turn would manifest into the firm's success in the global arena.

The study looks at the factors influencing Repatriation adjustment and satisfaction from the perspectives of Indian corporate.

Objective 3: To examine the relation between Repatriation and Reverse Cultural Shock

In contemporary times, the challenge of IHRM is to establish proper HR practices for selection of expatriate, managing expatriate, providing cross-cultural training for expatriate, providing allowance for expatriate and the lastly, managing repatriate. The study has analyzed factors relating to the holistic retaining of repatriates within the company which is very important in order to reduce costs and increase the companies' competitiveness. The principle reason for this study is to break down the repatriation process and to distinguish the primary determinants of fruitful repatriation. process. It has elaborated on the literature which conforms to the context of repatriates return to the home country, wherein there is a general assumption that the repatriate will be able to settle quickly into the community. As per research, this does not happen and many repatriates face re-entry shock or Reverse Culture Shock.

The study has empirically analyzed that Individual expectations before repatriation, during expatriation and after repatriation do not match the organizations' expectation in terms of various factors. The expectancy violation theory of met, unmet and under met expectations with regard to Repatriation adjustment of a Repatriate, form the core theories which explain the impediments and necessary requisites of Repatriation. Similar Anticipatory expectations of a Repatriate while overseas with respect to Repatriation back home, is a major hypothesis of the current study wherein the phenomenon of Reverse Culture Shock and its impact on such Anticipatory Adjustment has been statistically analyzed. The results of this study point out that there is a gap between the two states of anticipation of what lies in store for the Repatriate, on his return back home, from his perspective and what he/she actually experiences. This gap between expectation and reality forms the core area of Reverse Culture Shock context and this is where the bridging of gap is mandatory, as far as the strategic management aspect of International Human Resource Management and role of respective HR departments of organizations are concerned.

The study elucidated research literature from various other sources and elaborates the psychological, individual, sociological and organizational impediments that go into giving shape to the experience of Reverse Culture Shock during home country or in-country repatriation adjustment. This manifests itself in the context of adjusting to work, culture and general environment, wherein the antecedent variables comprise of Individual, Family, Job, Organizational and non work adjustment, as analyzed by Black, Gregorson and Mendenhall (1991).

The lack of understanding to integrate repatriates into the organization in terms of these factors and that of the knowledge and experience they gained during international assignment, is a factor that often leads many repatriates to quit the organization. Thus, the problem identified and analyzed is that of the impact of Reverse Culture Shock and repatriates difficulty in adjusting after an international assignment. Research on repatriation found that the culture shock of coming home is usually more difficult than the culture shock of going overseas Poor repatriation is costly, reduces the effective utilization of human resources, and often leads to a loss of talented The concept of "cross-cultural adjustment" began with earlier work on culture shock. The term Culture shock was introduced by an anthropologist, Dr. Kalervo Oberg (1960), to describe the trauma suffered by people when they are suddenly introduced to a culture that is very different from their own. In 1988, Befus, described culture shock as "an adjustment reaction syndrome caused by cumulative, multiple and interactive stress in the intellectual, behavioral, emotional and physiological levels of a person recently relocated to an unfamiliar culture and is characterized by a variety of symptoms of psychological shock. Individual expectations before repatriation, during expatriation and after repatriation do not match the organizations' expectation in terms of various factors at play respectively. This lack of understanding to integrate repatriates into the organization in terms of the knowledge and experience they gained during international assignment is a factor that often leads many repatriates to quit the organization. Thus, the problem identified for this study is the impact of Reverse Culture Shock and repatriates difficulty in adjusting after an international assignment.

Oberg (1960) described four stages of adjustment in the form of U-Curve theory of adjustment wherein the emotional stages of adjustment are divided into four stages, namely Honeymoon phase wherein the expatriates will have initial enthusiasm to interact with host nationals, lasting for a period of less than 2 months. The second stage corresponds to culture shock, when the individual has to cope up with the new culture on a day-day basis. In this stage, as per research, the individual lacks understanding the culture inhibits awareness of what is appropriate or inappropriate, behavior in the new cultural environment, resulting in confusion, frustration, tension and depression. The third stage is the adjustment phase characterized by increased ability to 'fit in' in the new culture. The last stage is the realignment or mastery phase wherein the adjustment is almost complete as possible, and anxiety is over. In short, individuals deputed on an international assignment first feel excited, then face problems and finally adjust over time to the new socio-cultural environment. Similarly the w-curve theory of adjustment by Gullahorn and Gullahorn (1963) argues that the cross-cultural adjustment process which led to a U-Curve can be applied to repatriation process as well. As per research, the longer one spends away from home; more changes can happen in home country. This leads to greater uncertainty upon return. The literature review of culture shock indicates that an individual may not function effectively during the period of cross-cultural adjustment. Research has indicated that repatriation adjustment was the strongest predictor of intent to leave the organization for Indian MNC's repatriates after repatriation. The repatriates who perceived a higher level of repatriation adjustment had a lower intent to leave.

While there is still much to be learned about repatriates' turnover intention and its antecedents and outcomes, this study has made an attempt to identify some major constructs which contribute for repatriate's adjustment in IT sector in India. Its results consolidate previous literature, by providing empirical evidence by analyzing the hypothesis that anticipatory repatriation adjustment is impacted by Reverse Culture Shock. The study is the first of its kind that identifies repatriates adjustment among corporate IT repatriates, within the ambit of the constructs of Competency Transfer,

Anticipatory Adjustment and Reverse Culture Shock, by simultaneously laying emphasis on the variables and requisites of Effective Repatriation. Subsequently, the conceptual framework of this study can be a manual for future examination in repatriates' adjustment.

The study provides evidence about the relationship between repatriation process and cultural re-entry impediments influencing repatriates' adjustment. The phenomenon of Reverse Culture Shock and its impact on such Anticipatory Adjustment has been statistically analyzed in the study. We proceeded with exploratory factor analysis to converge at two sub factors of Reverse Culture Shock (independent variable). Correlation between the factor groups of Reverse Culture Shock and average means of data from variables comprising Anticipatory Repatriate Adjustment was done, followed by Multiple Regression analysis between the identified factor groups of Reverse Culture Shock and the average mean of the statements of Anticipatory Repatriate Adjustment, to determine the plausible effect of Reverse Culture Shock on Anticipatory Adjustment. Multiple Regression analysis was applied between identified factors of Reverse Culture Shock, experienced by Repatriates and their Anticipatory Adjustment. The two identified factors of RCS are Alienation and Segregation (F1) and Loneliness and Anxiety (F2) and have been used as independent variables while the average mean of the statements of Anticipatory Adjustment (AA) has been taken as the dependent variable.

The multiple R value of 0.351 indicated a significant relationship between the variables and the adjusted R squared (0.112) suggested that the Reverse Culture Shock factors account for 11.2% of the variance in Anticipatory Repatriate Adjustment. $F = 11.050; p < 0.000$ was obtained after applying ANOVA wherein F Change is equal to 11.050. The results indicated that for each unit increase in independent variable, there is acceleration by 0.123 units (R-square = 0.123) and that the results are statistically significant, as depicted by β and t-values.

The standardized beta estimates for the factors of independent variable Reverse Culture Shock were significantly lesser than zero ($\beta = 0.074$ and -0.248, $p < 0.05$). The significant relation between dependent variable, Anticipatory Adjustment and independent variable, Reverse Culture Shock ($R = 0.351$, $p < 0.01$; $\beta = 0.074$ and -0.248, $p < 0.05$) demonstrated support for the fact that the factor has an impact on Anticipatory Adjustment.

This implies that the findings considerably denote that the criterion of the variables defining Anticipatory Adjustment, rely on the construct of Reverse Culture Shock and that the latter has an impact on the former whereby the Anticipatory Adjustment while overseas with respect to the foreseen adjustment in home country, gets negatively impacted. The results of the study point out that there is a gap between the two states of anticipation of what lies in store for the Repatriate, on his return back home, from his perspective and what he/she actually experiences. This gap between expectation and reality forms the core area of Reverse Culture Shock context and this is where the bridging of gap is mandatory, as far as the strategic management aspect and role of International Human Resource Management of organizations is concerned. This study has made an attempt to identify some major constructs which contribute for repatriate's adjustment in IT sector in India. Its results consolidate previous literature findings, by analyzing and providing empirical evidence to validate the hypothesis, that anticipatory repatriation adjustment is impacted by Reverse Culture Shock. The subsequent findings of the study point to a few routes in which global associations can expand their shots of holding important managers who come back from global assignments.

Objective 4: To undertake an impact assessment of repatriation on Competency Transfer, in the context of home country organizations

The domain of Competency Transfer of Expatriates on their Repatriation is one of the primary focus areas of the study. This factor comprises issues pertaining to the transfer of competencies of Repatriated employees per se. Research suggests that if international assignments were designed keeping in mind the different aspects of skill and acumen transfer principles, as elucidated in the variables under the construct of Competency Transfer, Repatriate talent would get optimized. It signifies that Repatriates who are able to make use of their overseas acquired acumen, resulting from their international experience tend to feel more satisfied with their career growth. Expatriation adds to revenue of the association as well as permits learning exchange that could have a positive effect on organizational learning and effectiveness. Therefore the strategies empowering expatriation, empowering superior performance, managing repatriation and enhancing employee satisfaction, should be well crafted and executed.

Therefore the objective was to undertake an impact assessment of repatriation process on Competency Transfer in the context of home country organizations. An analytical approach was undertaken for the said purpose by framing statements, pertaining to the items measuring Competency Transfer and Effective Repatriation. The former comprised of assessing Skill and acumen utilization, Knowledge sharing with co-workers, Skill transfer, Hold seminars, Organizational benefits, Career rewards, Decision making in International projects and Job position and Skill utilization. The item with the highest factor loading came out to be Job position and skill utilization.

The latter construct i.e. Effective Repatriation included nine items i.e. Repatriation training and benefits, Job position and hierarchy, Job description, Career progression ahead of colleagues, Role of communication in uncertainty elimination, Social status post repatriation, Acquired acumen and career development, No repatriation stress, Met expectations of Repatriation and Effects of changes in home country. The item with the highest factor loading was pertaining to acquired acumen and career development, implying the importance of knowledge acquisition and transfer and its further impact on career growth.

The objective was to undertake an impact assessment of repatriation process on Competency Transfer in the context of home country organizations. An analytical approach was undertaken for the said purpose by framing statements, pertaining to the items measuring Competency Transfer and Effective Repatriation. After subjecting the data collected to Exploratory factor analysis to converge at two sub factors of Effective Repatriation (independent variable), correlation and multiple linear regression analysis between the identified factor groups of Effective Repatriation and the average mean of the statements of Competency Transfer (dependent variable), to determine the effect of Effective Repatriation predictors on Competency Transfer outcomes.

A correlation test was done between the factor groups of independent variable, Effective Repatriation and Competency Transfer (dependent variable). The results indicated a positive and significant correlation between ER Factor 1 and CT avg (0.655) and significant correlation between ER Factor 2 and CT avg (0.645). Therefore the correlations for mean scale scores in indicated that the dependent variable Competency Transfer and the independent variable Effective Repatriation are moderately correlated as the correlation coefficients are less than 0.5 ($p < 0.5$) and therefore the significance of Effective Repatriation on the outcome of Competency Transfer.

Given the significance of this correlation, the multiple linear Regression analysis determined the effect of Effective Repatriation on Competency Transfer. The results indicated that for one unit increase in independent variable, there is an acceleration by 0.564 units (R Square = 0.564) and that the results are statistically significant as depicted by β and t-values. The multiple R value of 0.751 indicated a significant relationship between the variables and the adjusted R square (0.558) suggests that the Effective Repatriation factors account for 55.8% of the variance in Competency Transfer. $F = 101.394$; $p < 0.000$. The standardized beta estimates for the independent variable Competency Transfer were significantly lesser than zero. The statistical results demonstrated support for the fact that the factor has a positively significant relationship with Effective Repatriation.

This implies that the findings considerably denote that the criterion of the variables defining Competency Transfer, rely heavily on the construct of Effective Repatriation and that the latter has an impact on the former whereby Effective Repatriation considerably paves way for the transfer of the competencies of a Repatriate.

Objective 5: To propose Industry specific strategic model for Effective Repatriation

The paradigm of International Human Resource Management evolves slowly and therefore the gap between strategy implementation and practice as far as Repatriation Management is concerned can led to an unplanned and unstructured array of services and support mechanisms. Based on the aforementioned study, a model for Effective Repatriation is presented. Although many existing models address a broad scope of successfully managing repatriation and are based on several disciplines, sectors and country specific studies, this research study presents an integrated model based on data from a study of repatriation cases from Indian IT industry. This paper is also novel because in-depth studies of Repatriation adjustment and Effective Repatriation on Indian corporate sector are scarce and industry practitioners will always need to improve the retention of their international human resource.

Throughout the thesis the word "Effective Repatriation" is used. According to Cambridge Dictionary (2005) the term effective is "successful or achieving the results that you want". The word effective has arisen through the word efficiency which is defined as "when someone or something uses time and energy well, without wasting". When using these definitions together with the word repatriation one could understand what the authors of this thesis mean when referring to an Effective Repatriation. An Effective Repatriation is achieved when the repatriate is satisfied with the repatriation process because of a smooth assimilation to home nation and organization. Essentially expressed—Effective Repatriation is accomplished when all the aspects and people involved, have achieved the outcomes that they aspired for and consider the repatriation process as being successful. According to O'Sullivan (2002) a successful repatriation process can be defined as:

> One in which, upon return, the repatriate: gains access to a job which recognizes any newly acquired international competences, experiences minimal cross-culture readjustment difficulties; and reports low turnover intentions.

As proven by numerous research studies, a successful repatriate will normally be highly qualified, and have a suite of valuable skills, knowledge, and attributes and will have created many opportunities to grow, both culturally and technically. The growth of the individual should be accommodated in adaptation processes undertaken on return to their home country. Employees with exceptional results before their expatriate assignment ought to be supported and placed feasibly on return. This would give a feeling of job coherence, as well as would exhibit the worth the organization places on expatriate assignments. While this can guarantee that the expatriate procedure is fruitful, it turns out to be much more important that the arrival of this talent be tackled effectively. As repatriates, their services should be held secure and they must feel that their interests have been served by the organization. It is basically imperative that the repatriate sees a holistic support program to be in place International firms can influence their employees' commitment by developing effective strategies for repatriation. Such policies and practices have significantly reduced high repatriation turnover in many U.S. multinationals. While these may be seen as the province of larger, globalised firms, they should be prominent components of HRM processes for organizations dealing with repatriates. Devoid of a well structured policy formulation and implementation, the expatriates, may feel marginalized and concerned that they are being neglected whilst on assignment. The construction of an Effective Repatriation strategy is a key factor in retaining valued personnel.

Based on the present study, the effective protocol with respect to enabling Effective Repatriation should possibly comprise of training managers before they leave, supporting them while they are abroad, and acculturating them upon their return. The respective analysis of the constructs which have lead to conclusive findings as deliberated upon earlier are that an Effective Repatriation strategy should primarily comprise of integrated components, dealing with the principles and philosophy of the repatriation strategy, translated into a formal documented agreement, enacted through a supportive program. A re-entry evaluation is most compulsory in order to identify areas that require further consideration by the IHRM department. In order to achieve this, it is necessary to integrate Repatriation Management data into the Human Resource Management system and implement it effectively.

Various factors influence the repatriation of an individual when back in the home country. In accounting for these variables, models for developing an Effective Repatriation support process have been proposed by various researchers, keeping in view the peculiar nuances of the socio-political and cultural environment of a country, besides the general and organizational specifics, as applicable. Jassawalla et al. (2004) proposed that the quality of interaction with sponsors and perception of support upon return were found to be relevant. Reverse Culture Shock, repatriation adjustment, managing expectations, welcome home orientation, valuing international experience were also found to impact repatriation adjustment. Black gave an integrated model of International Adjustment which had three dimensions: Work Adjustment, Interaction Adjustment with foreign country nationals, Cultural Adjustment. The factors affecting expatriate adjustment are: Anticipatory Adjustment factors: Individual (pre-departure training, previous international experience etc.) and Organizational (effectiveness of selection procedure); In-country Adjustment factors: post departure adjustment issues.

As per the extensive analysis of Repatriation literature and that pertaining to the constructs of the present study and basis the findings of the statistical analysis of the data collected, in line with the proposed hypothesis, a specific strategic model for Effective Repatriation is proposed, wherein precisely a framework of four broad factors have an impact on Repatriation adjustment, from Organizational point of view, in eliminating the bottlenecks associated with Repatriation. These are; Policy development, which caters to recognizing the possible stress, repatriation may cause and the support of the company during time of transition.

6.4.1 Conceptual Model for Effective Repatriation

The respective analysis of the constructs which have lead to conclusive findings as deliberated upon earlier are that an Effective Repatriation strategy should primarily comprise of integrated components, dealing with the principles and philosophy of the repatriation strategy, translated into a formal documented agreement, enacted through a supportive program. A viable model of Effective Repatriation based on data from a study of repatriation cases from Indian IT industry has been proposed which caters to the feasibility criterion of the items predominantly at play, in making of the said constructs of the study. This study is also novel because in-depth studies of Repatriation adjustment and Effective Repatriation on Indian corporate sector are scarce and industry practitioners will always need to improve the retention of their international human resource through productive repatriation outcomes from organizational and employee perspective.

- It should start with strategic and pro-active expatriate selection, basis specific competence for dynamic global career management.
- Clarity of Job description and role in the organization, prior and post repatriation.
- Pre and post departure cross cultural assistance, career planning initiatives and a formalization of Repatriation policy, for before, during and after the assignment.

- Re-enforcement of realistic pre-return anticipatory repatriation expectations towards credible outcomes, with respect to work, interaction and general adjustment, via proper communication channels.
- Enabling appropriate management of expectations and transitions, thereby eliminating Reverse Culture Shock impediments related to Individual, Job, Organizational and Non-work variables, enabling adjustment.
- Reward, harness and optimize the acumen and skills acquired by the repatriate, while overseas, thereby enabling Competency Transfer through pro-active measures.
- Effective Repatriation outcomes which enhance human and intellectual capital of the organization would subsequently yield profitability from the company perspective. This implies lesser attrition rates, effective retention rates, optimization of human capital, lesser wastage, optimum return on investment and organizational growth.
- From an employee perspective, it would indicate smooth home country adjustment, lower anxiety and uncertainty, optimum assimilation in organizational and socio-cultural milieu, competency utilization and career progression, impetus to more global careers, knowledge exchange and transfer. So in totality, it comprises of work, interaction and general repatriation adjustment.
- Consequently, the evaluation of Repatriation experiences would yield a dynamic HR policy and tailor made strategies, towards employee and organizational profitability.

The model as proposed, basis the findings of the current study, identifies measures which can help formulate a pro-active repatriation strategy, determining measures that are favorable for Effective Repatriation and productive outcomes from organizational and employee perspective (Fig. 6.1).

6.5 Validity of Hypotheses Tested

Two main hypotheses were set up to be tested. The data generated from the present research has been used to test the hypotheses and validate the research assumptions herein.

H1: Reverse Culture Shock impacts Anticipatory Repatriation Adjustment

The hypothesis corresponds with the second objective of the present study, wherein correlation and multiple regression analysis were conducted on the collected data; to test whether Reverse Culture Shock impacts Repatriation adjustment.

An analytical approach was undertaken for the said purpose by framing statements, pertaining to the items measuring Anticipatory Adjustment before return home. Eight items comprised of statements covering the areas of Repatriation expenses, Smooth family adjustment, Role clarity and hierarchy of Job, Lesser role conflict, Improved social status, Favorable housing conditions, No effect of changes

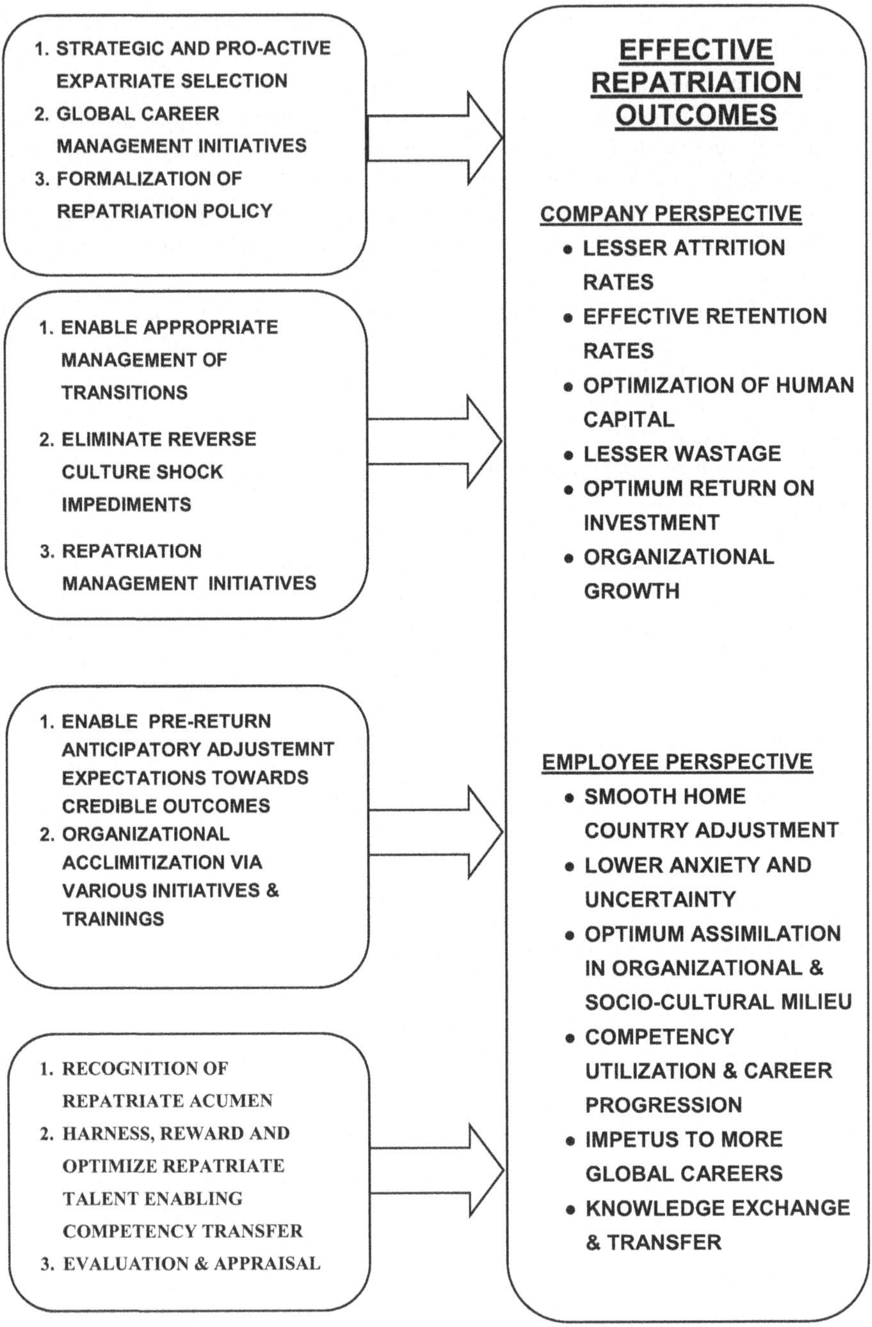

Fig. 6.1 Conceptual model for effective repatriation. *Source*: Basis the findings of the current study

back home, Smooth personal repatriation adjustment. The item with the highest factor loading turned out to be that pertaining to favorable housing conditions.

Reverse Culture Shock measurement scale comprised of nine items, covering statements pertaining to; Pre-Repatriation anxiety, Clash of expectations and reality, Problems in communication post Repatriation, Sense of frustration, Pre and Post Repatriation, Depression post Repatriation, Lonely and isolated post Repatriation, Alienation post Repatriation, Perceived Changes in values and attitudes of home country nationals post Repatriation, Missing foreign culture. The item with the highest factor loading turned out to be Pre Repatriation anxiety.

To test for the first hypothesis, correlation between the factors of Reverse Culture Shock and Anticipatory Repatriate Adjustment was done, followed by Multiple Regression analysis between the identified factor groups of Reverse Culture Shock (through Exploratory factor analysis) and the average mean of the statements of Anticipatory Repatriate Adjustment, to determine the effect of Reverse Culture Shock on Anticipatory Adjustment.

Items under construct of Reverse Culture Shock and Anticipatory Adjustment after applying factor analysis

The factor analysis findings for Anticipatory Adjustment revealed that the factor loadings range from 0.441 to 0.654, indicating that these values of the eight items of construct, pointed towards relatively fair importance of items in defining the dimensions of Anticipatory Adjustment. Total cumulative variance explained by the factor, Anticipatory Adjustment of Repatriates stood at 45.078%, with mean values being approximately equal and varying between 3.6500 and 4.1063. The total mean value stood at 3.8000 on a 5-point scale, and the total computed standard deviation of the items was 0.597, indicating the varying familiarity of the respondents to Anticipatory Adjustment, suggesting that all selected items significantly contributed towards the construct and that the subjects believed that the congruity between Anticipatory Adjustment and Repatriate Adjustment is important.

Factor analysis results for Reverse Culture Shock items revealed that the factor loadings range from 0.435 to 0.576 and significant at 0.05 levels, indicating high validity of the measurements of Reverse Culture Shock. Total cumulative variance explained by the factor Reverse Culture Shock experienced by Repatriates was 41.686%. The mean values varied between 2.2063 and 3.3188 and the total mean value stood at 2.77, which on a five point scale indicated the varying familiarity of the Repatriates to Reverse Culture Shock. The mean values of items of Reverse Culture Shock were approximately equal. The standard deviation was 0.730, which meant that the scores were spread away from the mean. On the whole, all selected items significantly contributed towards this construct, thereby suggesting, the subjects believe that the congruity between Reverse Culture Shock and Repatriate Adjustment is important.

Regression analysis between the two constructs (RCS and AA) and Demographic Variables

Regression analysis was done to examine the relationship between the two constructs as dependent variables and the demographic variables—Gender, Age, Marital status, Job profile, Length of stay overseas and Time elapsed since return from overseas, as independent variables. 16.7% of variability in the dependent variable, Anticipatory Adjustment was accounted for by all of the six demographic variables together. The β coefficients indicated that out of all six demographic variables, only the variable Age ($\beta = 0.242$; p-value = 0.000 has a significant influence on Anticipatory Adjustment).

11.1% of variability in the dependent variable, Reverse Culture Shock was accounted for by all of the six demographic variables together. The β coefficients indicated that out of all six demographic variables, the variable, Length of stay overseas ($\beta = 0.115$; p-value = 0.020) and Time elapsed since return to the home country ($\beta = 0.122$, and p-value = 0.002) had a significant influence on Reverse Culture Shock. All the respective demographics had varying degree of impact on the construct respectively.

Exploratory Factor Analysis to identify factors related to Reverse Culture Shock (Independent variable)

Exploratory Factor Analysis was used through SPSS (17 version) with Principal Component Analysis along with Varimax rotation for summarization of the total data falling under the construct of "Reverse Culture Shock", measured through the research instrument (Questionnaire), to arrive at meaningful factor groups. Such grouping of the statements of variables pertaining to the measurement of Reverse Culture Shock, in distinct areas of importance, yielded two factor groups; "Alienation and Segregation" comprising of four statements and "Loneliness and Anxiety" comprising of five statements, covering different dimensions of Reverse Culture Shock experienced by Repatriates. The implications for these factors are in relevance to the importance of these factors in impacting Anticipatory Repatriate Adjustment.

Correlation between Anticipatory Adjustment and factor groups of Reverse Culture Shock

Here RCS = Reverse Culture Shock, RCS Factor 1 = Alienation and Segregation, RCS Factor 2 = Loneliness and Anxiety, AA avg = Average of the means of variables defining Anticipatory Repatriate Adjustment. A correlation test was done between the said variables. The results, indicated negative but significant correlation between RCS Factor 1 and AA avg (−0.119) and negative but significant correlation between RCS Factor 2 and AA avg (−0.334). Therefore the correlations for mean scale scores indicated that the dependent variable Anticipatory Adjustment and the independent variable Reverse Culture Shock are moderately correlated as the correlation coefficients are less than 0.5 ($p < 0.5$). The results of the correlation matrix revealed that there exists a correlation among the variables of Reverse Culture Shock and those of Anticipatory Adjustment.

Impact assessment of Reverse Culture Shock on Anticipatory Adjustment through Multiple Regression Analysis

Multiple Regression analysis was applied between identified factor groups of Reverse Culture Shock, experienced by Repatriates and average mean of the variables defining Anticipatory Adjustment. The two identified factors of RCS are Alienation and segregation (F1) and Loneliness and Anxiety (F2) which were used as independent variables while the average mean of the statements of Anticipatory Adjustment (AA) has been taken as the dependent variable.

The multiple R-value of 0.351 indicated a significant relationship between the variables and the adjusted R squared (0.112) suggested that the Reverse Culture Shock factors account for 11.2% of the variance in Anticipatory Repatriate Adjustment. F = 11.050; $p < 0.000$ after applying ANOVA wherein F Change is equal to 11.050. The results indicated that for each unit increase in independent variable, there is acceleration by 0.123 units (R-square = 0.123).

The standardized beta estimates for the factors of independent variable Reverse Culture Shock were significantly lesser than zero ($\beta = 0.074$ and -0.248, $p < 0.05$). The significant relation between dependent variable, Anticipatory Adjustment and independent variable, Reverse Culture Shock ($R = 0.351, p < 0.01; \beta = \beta = 0.074$ and -0.248, $p < 0.05$), demonstrated support for the fact that the factor has a significant relationship with Anticipatory Adjustment.

7.4% of variability in Anticipatory Adjustment (AA) was accounted for by the first factor group RCS F1. Also 24.8% of variability in Anticipatory Adjustment (AA) was accounted for, by the second factor group (Loneliness and Anxiety).

The statistical results demonstrated support for the fact that factors Of Reverse Culture Shock (RCS) have an impact on Anticipatory Repatriate Adjustment (AA). This indicates that the impediments of Reverse Culture Shock are significantly predictive of an unfavorable impact on Anticipatory repatriation expectations of a repatriate and thereby hamper the process of smooth repatriation adjustment in home country. Therefore the results validate the Hypothesis H1, that Reverse Culture Shock impacts Anticipatory Repatriate Adjustment.

H2: Competency Transfer is significantly a function of Effective Repatriation

The hypothesis corresponds with the second objective of the present study, wherein a multiple regression analysis was performed in order to test whether Effective Repatriation is a significant predictor of Competency Transfer of a repatriate. Studies have reported that expatriates receive significant personal and self-development benefits from foreign assignments. They gain a broader, global perspective, skills, expertise and acumen on the firm's operations and these types of development can have positive career and organizational implications. It is desired that HR managers devise practices to better manage this critical resource (the expatriates) in order to maximize the wealth of knowledge and experience gained from international assignments. Performance development and management practices would have to be designed in such a way so as to capture the best of development opportunities. Specifically, it has been analyzed that the repatriates value more positively the extent to which their jobs have task variety and autonomy, provide opportunities for learning, allow them to apply their knowledge and give them enough responsibility. Researchers claim that the best part of international assignments is related to the new work experiences and learning they offer.

An analytical approach was undertaken for the said purpose by framing statements, pertaining to the items measuring Competency Transfer and Effective Repatriation. The former comprised of assessing Skill and acumen utilization, Knowledge sharing with co-workers, Skill transfer, Hold seminars, Organizational benefits, Career rewards, Decision making in International projects and Job position and Skill utilization. The item with the highest factor loading came out to be Job position and skill utilization.

The latter construct i.e. Effective Repatriation included nine items i.e. Repatriation training and benefits, Job position and hierarchy, Job description, Career progression ahead of colleagues, Role of communication in uncertainty elimination, Social status post repatriation, Acquired acumen and career development, No repatriation stress, Met expectations of Repatriation and Effects of changes in home country. The item with the highest factor loading was pertaining to acquired acumen and career development, implying the importance of knowledge acquisition and transfer and its further impact on career growth.

To test for the second hypothesis, we proceeded with exploratory factor analysis to converge at two sub factors of Effective Repatriation (independent variable). Correlation between the factor groups of Effective Repatriation and average means of data from variables comprising Competency Transfer of Repatriates was done, followed by Multiple Regression analysis between the identified factor groups of Effective Repatriation and the average mean of the statements of Competency Transfer, to determine the effect of Effective Repatriation predictors on Competency Transfer outcomes.

Items under construct of Competency Transfer and Effective Repatriation after applying Factor analysis

Factor analysis revealed that the item with the factor loadings for Competency Transfer ranges from 0.654 to 0.844 and significant at 0.05 levels. The total cumulative variance explained by the factor Competency Transfer of Repatriates was 57.332%. The mean values varied between 3.10 and 4.01 and the total mean value was 3.5414. The total computed standard deviation of the items was 0.828. On the whole, all selected items significantly contribute towards this construct, thereby suggesting that the subjects believe that the congruity between Competency Transfer and Repatriate Adjustment is important. The highest factor loading was 0.844 for the statement pertaining to Job position and skill utilization.

Factor analysis for Effective Repatriation revealed that the factor loadings range from 0.436 to 0.591. The communality for the nine statements was above 0.5, indicating moderate to high degree of linear association among the variables. Total cumulative variance explained by the factor Effective Repatriation of Repatriates was 49.932%. The mean values varied between 3.1313 and 3.7688 and the total mean value was 3.4139, on a 5-point Likert scale. The total computed standard deviation of the items stands at 0.728. On the whole, all selected items significantly contribute towards this construct, thereby suggesting, the subjects believe that the congruity between Effective Repatriation and Repatriate Adjustment is important.

Regression analysis between the two constructs (ER and CT) and Demographic Variables

Simple regression analysis was done to determine the impact assessment of Competency Transfer as dependent variable and the demographic variables (independent variables)—Gender, Age, Marital status, Job profile, Length of stay overseas and Time elapsed since return from overseas. 21.9% of variability in the dependent variable, Competency Transfer was accounted for by all of the six demographic variables together. The β coefficients indicated that out of all six demographic variables, Age ($\beta = 0.181$; p-value $= 0.024$) and Job profile as a Repatriate ($\beta = -0.355$, and p-value $= 0.001$) has a significant influence on Competency Transfer.

Likewise, Regression analysis was done to examine the relationship between Effective Repatriation as dependent variable and the demographic variables (independent variables). 14.1% of variability in the dependent variable, Effective Repatriation was accounted for by all of the six demographic variables together. The β coefficients indicated that out of all six demographic variables, the variables Age ($\beta = 0.160$; p-value $= 0.044$) and Job profile as a Repatriate ($\beta = 0.250$, and p-value $=$ 0.015) have a significant influence on Effective Repatriation.

Exploratory Factor Analysis to identify factors related to Effective Repatriation

The technique of Exploratory Factor Analysis was used through SPSS (17 version) with Principal Component Analysis along with Varimax rotation for summarization of the total data falling under the construct of "Effective Repatriation", measured through the research instrument (Questionnaire), to at two factor groups; "Career aspects"(five statements) and "Acumen, Communication, Changes and Repatriation Stress" (four statements), covering different dimensions of Effective Repatriation(ER) of Repatriates.

Multiple correlation between Competency Transfer and two factor groups of Effective Repatriation

Here ER = Effective Repatriation, ER Factor 1 = Career Description, Progression, Training role and Social Status post-Repatriation, ER Factor 2 = Acquired acumen and Job growth, Role of communication, Changes back home and Repatriation stress, CT avg = Average of the means of variables defining Competency Transfer. A correlation test was done between the said variables. The results show a positive and significant correlation between ER Factor 1 and CT avg (0.655) and negative but significant correlation between ER Factor 2 and CT avg (0.645). Therefore the correlations for mean scale scores indicated that the dependent variable Competency Transfer and the independent variable Effective Repatriation are moderately correlated as the correlation coefficients were less than 0.5 ($p < 0.5$). The results of the correlation matrix revealed that there is a positive correlation among the variables of Effective Repatriation and those of Competency Transfer.

Impact assessment of Effective Repatriation on Competency Transfer

A multiple regression analysis was performed between identified factors of Effective Repatriation(ER), of Repatriates and their Competency Transfer, in order to test whether Competency Transfer is significantly a function of Effective Repatriation, wherein Competency Transfer remains the dependent variable and Effective Repatriation is the independent variable.

The two identified factors of ER are "Career Aspects" and "Acumen, Communication, Changes and Repatriation Stress", and have been used as independent variables while the average mean of the statements of Competency Transfer (CT) has been taken as the dependent variable. The multiple R value of 0.751 indicated a significant relationship between the variables and the adjusted R square (0.558) suggested that the Effective Repatriation factors accounted for 55.8% of the variance in Competency Transfer.

The standardized beta estimates for the dependent variable Competency Transfer were significantly lesser than zero ($\beta = 0.276$ and 0.373, $p < 0.05$). The statistical results demonstrated support for the fact that the Competency Transfer is significantly a function of Effective Repatriation. After applying ANOVA, F change is equal to 101.394 ($p < 0.000$).

27.6% variability in CT (Competency Transfer), the dependent variable was accounted for, by the first factor group variables i.e. ER F1. Also 37.3% of variability in CT was accounted for; by the second factor group i.e. ER F2. The statistical results demonstrated support for the fact that factors of Effective Repatriation (ER) have an impact on Competency Transfer (CT). This implies that the findings considerably denote that the criterion of the variables defining Competency Transfer, rely heavily on the construct of Effective Repatriation and that the latter has an impact on the former whereby Effective Repatriation considerably paves way for the transfer of the competencies of a Repatriate.

This indicates that strengthening the parameters of Effective Repatriation is significantly predictive of an increase in Competency Transfer. When a Repatriate is effectively repatriated, his competency and knowledge is transferred and optimized, thereby resulting in his career advancement, Organizational adjustment, as compared to the scenario when vice versa happens and may lead to repatriate dissatisfaction and attrition. Therefore this validates the Hypothesis H2; Competency Transfer is significantly a function of Effective Repatriation.

Chapter 7
Policy Recommendations for Better Future

7.1 Findings of the Study

The present study has taken a profound view of Repatriation adjustment and its antecedents in terms of four constructs which may impact it, namely Competency Transfer viability of Repatriates, Effective Repatriation process of Job assignees on their return back home, the anticipatory expectations regarding Repatriation and the impact of Reverse Culture Shock on the whole process. It was hypothesized that a Repatriates return back home is the function of his assimilation in the milieu back home, wherein his anticipation of repatriation may get impacted by Reverse Culture Shock where home may not feel and appear like home. Similarly his acumen and expertise assimilated while on the foreign sojourn is likely to get transferred as a consequence of effectively getting repatriated back in the home country. The results of the study revealed the met and unmet expectations of repatriates in the backdrop of their repatriation where the effectiveness of Repatriation Management by the home organizations and the socio-cultural environs, have a direct impact on the success of repatriation.

An exploratory, empirical study to obtain a fresh insight into Repatriation Experiences and Repatriate Adjustment of employees from India, who had been on international assignments, has been done, within the ambit of four constructs: Competency Transfer and its significant relationship with Effective Repatriation; Anticipatory Adjustment and it's the impact of Reverse Culture Shock on it. The findings authenticated the respective hypotheses of the study and thus gave the following insights:

- The Repatriate expectations about the assimilation of knowledge, skills and abilities gained in the host country and its optimization back in the home country, moderate the relationship between Effective Repatriation in the home country and Repatriate Adjustment, such that the transfer and utilization of such acumen of

M. B. Nag and F. Ahmad Malik, *Repatriation Management and Competency Transfer in a Culturally Dynamic World*,
https://doi.org/10.1007/978-981-19-7350-5_7

respondents can decrease their dissatisfaction and eventually lead to their adjustment within the organization. Therefore authenticating that Competency Transfer is significantly a function of Effective Repatriation. Global organizations that set realistic expectations about re-entry career opportunities for repatriates, facilitate knowledge transfer after repatriation, and adequately respond to boundary less career ambitions of repatriates, by realization of their anticipatory expectations regarding home, which can eventually facilitate Repatriate Adjustment, and hence may reduce repatriate turnover intention and attrition.

- The Anticipatory expectations of the Repatriates; Pre-departure, with respect to Repatriation back home and their consequent Pre-return adjustment, to such expectations, is impacted by the Reverse Culture phenomenon which results in a gap between the anticipated expectations and reality of Repatriation and eventual adjustment. The overseas anticipated adjustment and in-home adjustment therefore show varying degree of difference, calling for a closure of such a gap by elimination of Reverse Culture Shock impediments and thereby guarantee overall Repatriate Adjustment.
- As more companies expand their business overseas, the need to send employees on international assignments and retaining them has increased, making the process of discovering the reasons for these failures, increasingly important. Factors such as career planning, financial issues, family problems, re-establishing back into the firm and country, lack of training, stress, duration of the assignment, age, previous culture experience and effect on family life can be offset by holistic analysis and strategic planning by MNCs
- It was hypothesized that a Repatriates' return is the function of his assimilation in the milieu back home, wherein his anticipation of repatriation may get impacted by Reverse Culture Shock where home may not feel and appear like home. Similarly his acumen and expertise assimilated while on the foreign sojourn is likely to get transferred as a consequence of effectively getting repatriated back in the home country. The results of the study revealed the met and unmet expectations of repatriates in the backdrop of their repatriation, where the effectiveness of Repatriation Management by the home organizations and the socio-cultural environs, have a direct impact on the success of repatriation. All this needs to be looked into, strategically by the policy makers.
- Basis the framework of constructs and the findings of the study, a viable model of Effective Repatriation has been proposed which caters to the feasibility criterion of the items predominantly at play, in making of the said constructs. This involves a phased out strategic planning and implementation in various stages of Repatriation, starting from the selection of expatriates, during the stay, and pre-return support from home organizations and post departure and post return facilitation of Repatriation adjustment.
- Strategic and pro-active expatriate selection and global career management (pre and post departure cross cultural training), re-enforcement and enabling pre-return anticipatory adjustment expectations towards credible outcomes, enabling appropriate management of transitions and elimination of Reverse Culture Shock

impediments, leads to adjustment and therefore harnesses, rewards and optimizes repatriate talent, thus enabling Competency Transfer.

- Effective Repatriation outcomes would subsequently comprise of two facets; A Company perspective which would encompass lesser attrition rates, effective retention rates, optimization of human capital, lesser wastage, optimum return on investment and organizational growth.
- From an Employee perspective, smooth home country adjustment, lower anxiety and uncertainty, optimum assimilation in organizational and socio-cultural milieu, competency utilization, career progression, impetus to more global careers, knowledge exchange and transfer, is necessitated, covering work, interaction and general repatriation adjustment.
- The findings, basis the study of Indian IT Sector Repatriates substantiates and validates Black et al.'s framework of Repatriation adjustment, implying that the variables at play in the Indian context remain the same. These variables are predominantly related to pre-return adjustment (Communication, Sponsors, visits to home country, pre-return training and orientation) and pre-return expectations pertaining to work, interaction and general environment.
- The variables at play, as far as Post return adjustment is concerned, also stand validated in the context of study within the parameters of the constructs taken. This pertains to Individual, Job (Job discretion, job clarity, promotion and utilization of international skills), Organizational (clear repatriation process, financial compensations and post arrival orientation/training) and non work variables (social status, housing conditions and family adjustment). A holistic interpretation and subsequent policy formulation is called for.
- IT professionals form a large chunk of the total knowledge workers working globally their motivations are largely guided by their perceptions of organizational support, both intrinsically challenging jobs and a certain degree of self management on the job. This in turn implies that HR professionals need to chart out tangible career paths, devise efficient performance management systems and challenging job designs so as to motivate employee satisfaction and performance, as an Expatriate and a Repatriate.
- The role of HR departments in smoothening out the crevices, if any, and removing the impediments in the way of successful repatriation, may have a great influence. The results show that Reverse Culture Shock acts as a negative moderator between Repatriation and acculturation of repatriates, back home, revealing that repatriates go through a cyclical phase of adjustment back home, wherein the initial phase comprises of euphoria of return back home, the second phase is of unmet anticipatory expectations or dejection, the third phase is of recovery and reconciliation with reality and the final phase marks the final adjustment of a repatriate. Given the plethora of variables at play, HR initiatives can eliminate these impediments of Repatriate Adjustment.
- The results provide an understanding of Repatriate' psychology and the variables impacting its association with adjustment back home, starting from the phase of expatriation. It may give the practitioners some ideas in understanding the

concept of Repatriation Management in varied hues, namely socio-cultural, organizational, psychological, personal and family related and individual constructs. The four constructs under study encompass in totality, the various domains which may go into the understanding of the concept. This is validated by the literature review elucidated earlier.
- The results accentuate the view that Competency Transfer is a function of Effective Repatriation. An increase in the effectiveness of repatriation significantly increases the Competency Transfer effectiveness via an interaction effect. This provides fresh insights into Repatriation Management as far as knowledge transfer of Repatriates is concerned, as applicable in Indian corporate context, which had not been researched upon earlier, under the aegis of the studied constructs.
- Organizations need to pay attention to perceived Repatriation Concerns, so as to ensure that the individual does not perceive Repatriation to be a detriment to career development. Organizations need to be proactive about re-absorption of returnees and subsequent advancement so as to create positive perceptions in the minds of those working on international. Repatriation programs need to encourage and train employees to practice proactive career planning behaviors such as taking inventory of the skills developed during their international assignment setting goals based on how they want to utilize these competencies in their home country environment.

It is pertinent to point that the present study is based on corporate employees in IT sector in India. Given the global management practices in Indian IT firms, exposure to global media in all aspects of life and global influences, it calls for a thorough research on the optimization status of such talent who otherwise can fall prey to reverse brain drain.

7.2 Research Contributions

Scope for further research and the limitations which exist in the current literature concerned with repatriation in Human Resource Management have given insight and importance to the current research, in constructing this dissertation. It highlights the importance of Repatriation Management internationally to set up programs which can fulfill the gap on the study of repatriates and the issues facing repatriate adjustment.

In this study, an attempt has been made to obtain a comprehensive understanding of the variables impacting and influencing Repatriation adjustment. International assignments are the pivot of international HR which makes it a fact that they often are a great concern to many corporations who make big investments in overseas assignments. As more companies expand their business overseas, the need to send employees on international assignments and retaining them has increased, making the process of discovering the reasons for these failures, increasingly important. The study provides an insight into the following realms:

- Factors such as career planning, financial issues, family problems, re-establishing back into the firm and country, lack of training, stress, duration of the assignment, age, previous culture experience and effect on family life, can be offset by holistic analysis and strategic planning by MNCs.
- The repatriation and retention process can be a valuable human resource activity that could assist in significantly improving an organization's ability to compete in the global arena. The skills and experience retained through this process can play a major role in ensuring the future success of the multinationals, which properly use it. The repatriation and retention process should be done as a formal program. Provisions for using the skills that the expatriate developed in the foreign environment are a highlighted area of study. Thus, a proactive International Human Resource Management program can provide very valuable assistance in over-watching this process and implementing the requisite recommendations as proposed by the study. It augurs well for Indian multinationals to conform to the above mentioned suggestions, so that the repatriation and retention process can be executed in a manner that benefits the organization, the expatriate-repatriate, and the family.
- The result of this thesis proposes that the expectations and the process of adjustment have a significant impact on how the repatriation process is perceived. A majority of the factors identified above have previously been acknowledged by researchers. Nevertheless, this study highlights an inter-linkage of factors in Indian IT context. The present research provides theoretical contribution by filling the gaps in the existing literature, wherein no study in Indian corporate context has been undertaken in an elaborate manner, given the constructs of Competency Transfer, Effective Repatriation, Anticipatory Adjustment and Reverse Culture Shock; that define Repatriation adjustment. The study can provide insights for further research in this area and help the Human Resource Departments functioning internationally; to formulate policies that harness Repatriate talent and discourage the wastage that takes place as a result of widespread attrition of Repatriates due to dissatisfaction with Repatriation. The study is based on relevant theories and empirical information. The aim is to analyze the repatriation scenario in IT industry and from the analysis, intends to come up with conclusions on how to improve the repatriation program and thus help in minimizing the repatriation turnover.
- This paper describes several constructs pertaining to Repatriation Management which may impact the repatriate support practices offered by organizations prior to the departure, during the assignment abroad and upon return from international assignment. Thus, the main aim of this paper is to explore the reasons behind successful repatriation, after International Assignment.
- The proposed model provides the HR Departments with a framework to understand the concept of fruitful Repatriation, given the various variables that come into play while the process takes place. HR departments of Corporate India, Expatriates and Repatriates can use the suggestions and findings of the study to rationalize the use and effectiveness of Repatriate talent, in a highly competitive global market.

- Four constructs with varied measurable items have been analyzed which pave way for certain recommendations involving pre-planning, communication, re-positioning basis acumen acquired, Competency Transfer, during all stages of international assignment, aimed towards Effective Repatriation, in terms of both improving the current repatriation process as well as creating a sense of accomplishment for the Repatriate who returns back. The repatriation and retention process should be done as a formal program. Thus, a proactive international human resource management program can provide very valuable assistance in over watching this process and implementing the requisite recommendations as proposed by the study. It augurs well for Indian multinationals to conform to the above mention suggestions, so that the repatriation and retention process can be executed in a manner that benefits the organization, the expatriate-repatriate, and the family.
- The statistical results demonstrated support for the fact that Reverse Culture Shock impacts Anticipatory Repatriate Adjustment. This indicates that increase in the level of Reverse Culture Shock (low to high) was significantly predictive of a decrease in Anticipatory Adjustment of a Repatriate in the selected domain. This indicates that the impediments of Reverse Culture Shock are predictive of an unfavorable impact on Anticipatory Repatriate expectations of a Repatriate and thereby hamper the process of smooth Repatriate Adjustment in the home country.
- The statistical results demonstrated support for the fact that Competency Transfer is a statistically significant function of Effective Repatriation. This indicates that increase in the level of Effective Repatriation (low to high) was significantly predictive of an increase in Competency Transfer of a Repatriate. This indicates that when a Repatriate is effectively repatriated, his Competency is transferred and optimized; thereby resulting in his career advancement, Organizational adjustment, as compared to the scenario when vice versa happens and may lead to repatriate dissatisfaction and attrition.
- These results are likely to help academicians and researchers to collect more information about various other sub factors and variables at play and generate an understanding for policy formulation at macro level in a rapidly developing Globalized India. Given the scenario of macro level efforts being made at Central and state levels to accelerate the country towards a more globalized presence and the "Make in India" campaign taking birth; there's an urgency of tapping the indigenous and Repatriated talent to check both Brain drain and accelerate Reverse brain drain.
- This study suggests the practical issues, which a Repatriate may face upon return, and which can be a factor of significance studying the perception of a repatriate about the repatriation process, thereby providing a psychological perspective from a Repatriates point of view. It combines the constructs of culture and psychology, interwoven in the broader gamut of corporate repatriation of employees and that such interconnectedness thereof was likely to generate unfavorable attitudes towards the company, and ultimately, lead to turnover.

- The findings of the study highlighted that there was a significant relationship between being effectively repatriated and transfer of their acumen and competency at large, which may further aid career growth, employee and organizational development. In particular, the practices imperative to favorable repatriation and organizational and cultural support along with individual traits have significantly facilitated retaining repatriates.

Therefore keeping in view these paradigms, the study has highlighted the interaction equations between variables at play and the underlying impediments that may obstruct smooth Repatriation of Corporate employees. An improved repatriation process delivers a message that the company appreciates international experience, and therefore, strives to compete successfully in the global market. In the end, an Effective Repatriation process will probably result in a higher employee demand for international assignments.

7.3 Managerial Implications and Recommendations for Practitioners

The present study has tried to provide analytical details of a Repatriates understanding and feedback to Repatriation, as he/she understands it from personal and organizational perspective. This for sure has implications for practitioners and repatriates alike, from a management point of view.

This study attempts to identify factors that impact Repatriate satisfaction in the Indian IT Industry. It proposed and tested how the different dimensions of Repatriate experience are linked to Adjustment back in the home country. The different dimensions of Repatriate experience were derived from existing literature and studies focusing on those dimensions. Indian IT professionals were surveyed on their experiences post overseas assignment and factors such as Anticipatory Adjustment, Competency Transfer, Effective Repatriation and Reverse Culture Shock were identified through exploratory factor analysis. There exists limited research on Indian Corporate Repatriate experiences after International assignments and hence the study can be a future reference tool in the said area with the applicable constructs, from strategic HR managerial perspective.

The general issues addressed in this study are what factors lead to satisfaction (or dissatisfaction) among Repatriates and what the Corporate HR can do to enhance their employees' In-country adjustment and commitment to their jobs and assignments, thereby preventing attrition post Repatriation. A common theme that underlies all factors derived in the study is the need for continued support from the individuals' organization. From efficient selection procedures (person-job and person-culture fit), pre-departure training and orientation, career advancement opportunities and re induction upon return-A gamut of responsibility lies with the organizations sending these professionals on assignments.

The study provides several important managerial implications for HR Departments of corporate in the context of Effective Repatriation of Expatriates on their return back home. The broad constructs of the study need a relook as far as their credibility in terms of managerial context is concerned. From the practitioner's point of view, the research has implications as far as the implementation of a career plan that should utilize the experiences and knowledge of the repatriates that they have gained during the international assignment, is concerned. Organizations while making up their strategy should bear in mind that there should be a strong bond for the repatriates with their home countries through communication with them before, during and after the assignment. MNC can build a pool of international management expertise that can advise and incorporate the features to their management of repatriation and subsequently International assignments should be a development tool to build the ranks of experienced international managers.

- Organizations while making up their strategy should bear in mind that there should be a strong bond for the repatriates with their home countries through contacting them before, during and after the assignment, thereby calling for a feasible induction program for the repatriates, as to give them a newer feel of the home country. This enables an awareness of the corporate culture for the assignee before hand, as to prevent him from a cultural shock. Induction programs for the repatriates, as to give them a newer feel of the home country and assistance for job prospects on his re-entry are a pre requisite for satisfactory Repatriation adjustment. MNC should implement a career plan that should utilize the experiences and knowledge of the repatriates that they have gained during the international assignment and provide befitting job prospects to the repatriate on his re-entry.
- MNC should build a pool of international management expertise that can advise and incorporate the features to their management of repatriation. Therefore, International assignments should be a development tool to build the ranks of experienced international managers. Practitioners who had been through a repatriation process should see the motives of the managers to go abroad and the key motive should be organization and streamlining of a successful repatriation program for the international assignee.
- Fringe benefits and bonus packages as a part of the package for the repatriate, to eliminate repatriation related stress and tax reduction on the repatriates financial earnings so that he doesn't feel alienated and can work as a good team player. This facilitates and gives impetus to confidence building measures to retain talent and discourage attrition.
- Findings of the study indicate that there should be a conscious effort on part of Organizations to impart awareness of different cultures around the globe with revised cross cultural adaptation variables at play. Research and consequent policy implementation should be carried out on the effects such as stress, cultural shock, and reduction of autonomy and family problems on managers and how to offset them. There should be research on how to give awareness of different cultures around the globe. Culture shock and Reverse Culture Shock of Repatriates need a deeper insight and focus.

- Cross cultural training initiatives, realistic job previews and subsequent socialization practices are of pivotal importance in this regard. Given that IT projects cover a large range of technical knowhow and roles ranging from developers, analysts, system architectures and managers; paying heed to their adjustment back home with respect to in-country and organizational adjustment is important. Feelings of stress in a new work environment, ability to decode the communication (verbal and non-verbal) and social etiquette and acclimatizing oneself with the way work is done in terms of interpersonal and team protocols comprise this set of factors which impact satisfaction, a Repatriate feels on his return back home. Managers need to formulate policies to close this gap.
- Effective Repatriation training and socialization methods need to be employed by parent organizations. It is imperative that not just technical considerations but also an employees' willingness and motivation to relocate, albeit temporarily, are not ignored.
- The study can help further the research that has been done on the topic, in the hope of bettering the understanding of the topic and reinforce the previous research on Repatriation. Practitioners can take cues from the findings of the study and use it to further their decision making process when looking for practical measures for policy formulation as far as Repatriation Management is concerned. Effective Repatriation policy that addresses the issues pertaining to cross cultural adjustment both in the workplace and in daily life will go a long way towards optimization of manpower resource and elimination of wastage.

Therefore the Corporate in general and Repatriates in particular will benefit from this research in understanding the impediments that accelerate the attrition rates of Repatriates due to faulty Repatriation and under utilization and poor optimization of Competencies gained overseas.

7.4 Limitations

Given the scale and dimensions of the study, there were bound to be certain limitations from various perspectives. These were generally associated with the inaccessibility of the huge database and the busy schedule of the IT employees. Questionnaire responses got delayed on various accounts as a result of this and so did the data collection and analysis.

- The inaccessibility of the huge database and the busy schedule of the IT employees got the Questionnaire responses delayed on various accounts.
- The random perception of some employees that maybe they would divulge too much information of secretive nature, was a bottleneck. On the contrary, many perceived the information given, to be a reflection on their personal lives or bitter experiences, something everyone is not willing to reveal. Employees at times are also reluctant to share their information because of privacy reasons and may therefore give false information.

- Sometimes the recall value of Repatriates related to their Repatriation experiences suffered on account of the paucity of time. Data presented by the participants represents a retrospective account of the experience of repatriation.
- Participants had been repatriated for varying amounts of time. This raises speculation that differing conclusions may have resulted if analysis of the experiences of participants occurs throughout the process of repatriation. Therefore, repatriation as an evolving process needs longitudinal investigation.
- The study is based on primary data collected from a relatively small sample of IT professionals who have repatriated back to India. Given that, culture, immigration rules and other factors relating to expatriate experiences would differ between different countries of assignment, the factors identified in this study may not be universally applicable to all expatriates across nations.

7.5 Future Research Implications

Despite the limitations of this study, this study does give human resource professionals an idea of what focus points should be looked at while designing jobs and related HR practices for expatriates so as to minimize expatriate failure thereby having a positive impact on profitability and organizational success. An empirical study has been developed which needs to be delved deeper along similar or alternate lines with the theoretical and managerial implications, for policy makers and stake holders.

- Expatriation which marks the initial phase of the circular process of Repatriation adjustment, calls for a more in-depth research into the variables which go into the policy formulation of Expatriation programs, as it exists in Corporate India.
- Research could be extended to repatriate experiences in different international locations and across different sectors e.g. manufacturing, health and service oriented separately. Comparative studies can also be made between the manufacturing and service oriented industries across the large number of cities from India.
- Reverse brain drain studies can be a derivative of Repatriation Management domain and its research can hold immense value in the context of manpower utilization and optimization, with minimum possible wastage.
- Socio cultural context of the Repatriation Adjustment can be a significant area of study. Repatriated labor class from India and Repatriation from and in conflict zones needs to be studied upon. This can have major implications for various policy and law frameworks.
- It is evident from research that factors like spouse support, family adjustment, spouse career, children's education, support from other family members also impacts Repatriate Adjustment. Results of such studies, as the reciprocal effects of adjustment between family members may contribute to the understanding of this complex process.

- Further, it would also be interesting to look upon international assignments and the repatriation process from a broader and more international perspective. For instance, how has the globalization affected the management of international assignments and what impact this change might change have on the individual expatriate.

The number of employees working internationally will continue to grow, and so must our understanding of the repatriation experience, and ways to support employees during their transition home. This study could function as an extension of previous research providing an in-depth understanding of the concept under scanner.

GPSR Compliance

The European Union's (EU) General Product Safety Regulation (GPSR) is a set of rules that requires consumer products to be safe and our obligations to ensure this.

If you have any concerns about our products, you can contact us on ProductSafety@springernature.com

In case Publisher is established outside the EU, the EU authorized representative is:

Springer Nature Customer Service Center GmbH
Europaplatz 3
69115 Heidelberg, Germany

The manufacturer's authorised representative in the EU is Springer Nature Customer Service Centre GmbH, Europaplatz 3, 69115 Heidelberg, Germany. If you have any concerns regarding our products, please contact ProductSafety@springernature.com

Printed and bound by CPI Group (UK) Ltd, Croydon, CR0 4YY
30/09/2026
02225320-0002